Springer Theses

Recognizing Outstanding Ph.D. Research

Aims and Scope

The series "Springer Theses" brings together a selection of the very best Ph.D. theses from around the world and across the physical sciences. Nominated and endorsed by two recognized specialists, each published volume has been selected for its scientific excellence and the high impact of its contents for the pertinent field of research. For greater accessibility to non-specialists, the published versions include an extended introduction, as well as a foreword by the student's supervisor explaining the special relevance of the work for the field. As a whole, the series will provide a valuable resource both for newcomers to the research fields described, and for other scientists seeking detailed background information on special questions. Finally, it provides an accredited documentation of the valuable contributions made by today's younger generation of scientists.

Theses are accepted into the series by invited nomination only and must fulfill all of the following criteria

- They must be written in good English.
- The topic should fall within the confines of Chemistry, Physics, Earth Sciences, Engineering and related interdisciplinary fields such as Materials, Nanoscience, Chemical Engineering, Complex Systems and Biophysics.
- The work reported in the thesis must represent a significant scientific advance.
- If the thesis includes previously published material, permission to reproduce this must be gained from the respective copyright holder.
- They must have been examined and passed during the 12 months prior to nomination.
- Each thesis should include a foreword by the supervisor outlining the significance of its content.
- The theses should have a clearly defined structure including an introduction accessible to scientists not expert in that particular field.

More information about this series at http://www.springer.com/series/8790

Hideyuki Hotta

Thermal Convection, Magnetic Field, and Differential Rotation in Solar-type Stars

Doctoral Thesis accepted by
The University of Tokyo, Tokyo, Japan

 Springer

Author (Present Address)
Dr. Hideyuki Hotta
HAO/NCAR
Boulder, CO
USA

Supervisor
Assoc. Prof. Takaaki Yokoyama
The University of Tokyo
Tokyo
Japan

ISSN 2190-5053
Springer Theses
ISBN 978-4-431-56258-0
DOI 10.1007/978-4-431-55399-1

ISSN 2190-5061   (electronic)

ISBN 978-4-431-55399-1   (eBook)

Springer Tokyo Heidelberg New York Dordrecht London

Springer Japan KK is part of Springer Science+Business Media (www.springer.com)

Parts of this thesis have been published in the following journal articles:

H. Hotta, M. Rempel, and T. Yokoyama, 2014, ApJ, 786, 24
H. Hotta, M. Rempel, and T. Yokoyama, 2015, ApJ, 798, 51

Supervisor's Foreword

This thesis describes the studies on the solar interior where turbulent thermal convection plays an important role. Dr. Hideyuki Hotta solved, for the first time, one of the long-standing issues in solar physics, i.e., the maintenance mechanism of the solar differential rotation in the near-surface shear layer. He attacked this problem with his newly developed approach, the reduced speed of sound technique, which enabled him to study the surface and deep solar layers in a self-consistent manner. This technique also made it possible to achieve an unprecedented performance in the solar convection simulations for use of massively parallel supercomputers such as the RIKEN K system. It was found that the turbulence and the mean flows such as the differential rotation and the meridional circulation mutually interact with each other to maintain the flow structures in the sun. Recent observations by helioseismology support his proposed theoretical mechanism. He also addressed the generation of the magnetic field in such turbulent convective motions, which is an important step forward for the solar cyclic dynamo research.

Tokyo, Japan, December 2014 Assoc. Prof. Takaaki Yokoyama

Acknowledgments

I would like to express my gratitude to my supervisor Takaaki Yokoyama for his contribution to my education in the last 6 years. I would also like to thank Matthias Rempel of the High Altitude Observatory (HAO), USA, for his assistance during my stay in Boulder and his insightful advice. I greatly appreciate the help of HAO scientists Mark Miesch, Nick Featherstone, and Yuhong Fan. I acknowledge the members of the Yokoyama lab Yusuke Iida, Naomasa Kitagawa, Shin Toriumi, Yuki Matsui, Haruhisa Iijima, Takafumi Kaneko, Shyuoyang Wang, Shunya Kono, and Tetsuya Nasuda, and the members in the STP group at the University of Tokyo for helpful discussions in the seminars and colloquium. The numerical calculations in this thesis were carried out in the K-computer and FX10. My work was supported by a JSPS fellowship. My stay at the HAO and the Max Planck Institute were supported by the JSPS study abroad program. Finally, I would like to express my deepest thanks to my parents and other family members for their continuous support and encouragement.

Contents

Chapter 1
General Introduction

Abstract This chapter serves as the introduction. In the sun, a static radiative zone and turbulent convection zone exist. The convection zone is filled with turbulent thermal convection that transports the energy and angular momentum. On account of the significant change in the pressure scale height from the base of the convection zone (60,000 km) to the photosphere (300 km), the convection size drastically changes along the radius. This change is the central issue of this thesis. The energy transport is one of the most important factors in solar stratification models. The angular momentum transport generates mean flows, such as differential rotation and meridional flow. The drastic change in the time scale of the convection causes the shear layer near the surface. In the introduction, we discuss the current understanding on the basis of the previous studies and the remaining problems.

Keywords Solar interior · Solar large-scale flow · Reynolds stress

1.1 Solar Structure and Mean Flow

1.1.1 Solar Structure

The sun consists of the radiative zone (from the center to $0.715R_\odot$) and the convection zone (from $0.715R_\odot$ to the surface), where $R_\odot$ is the solar radius ($= 6.960 \times 10^{10}$ cm). Figure 1.1 shows the distribution of physical variables, i.e., gravitational acceleration, density etc., which are taken from the solar standard model (Model S: Christensen-Dalsgaard et al. 1996). Solar luminosity can be estimated from observations ($L_\odot = 3.84 \times 10^{33}$ erg s^{-1}). The other variables in the solar interior are calculated by using the observed luminosity and surface abundance under the assumptions of (1) spherical symmetricity (one dimension), (2) hydrostatic equilibrium, i.e., the balance of the pressure gradient and the gravity force, and (3) thermal equilibrium, i.e., the energy balance between nuclear fusion and transport. The effects of rotation and magnetic field are not included in the solar standard model. Because the plasma temperature exceeds 10^7 K near the core, energy is continuously generated by nuclear fusion.

© Springer Japan 2015

H. Hotta, *Thermal Convection, Magnetic Field, and Differential Rotation in Solar-type Stars*, Springer Theses, DOI 10.1007/978-4-431-55399-1_1

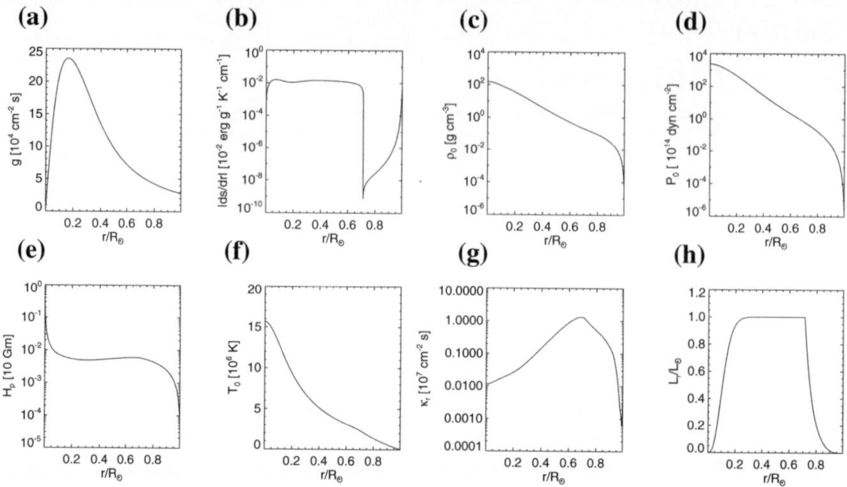

Fig. 1.1 Various physical quantities in the solar interior from the center to the surface estimated with the solar standard model: **a** gravitational acceleration, **b** absolute value of the specific entropy gradient, **c** density, **d** gas pressure, **e** pressure scale height, **f** temperature, **g** radiative diffusivity, **h** radiative luminosity. The values are obtained from Christensen-Dalsgaard et al. (1996) and his website

In the radiative zone, this energy is transported by efficient radiation. From $0.2R_\odot$ to $0.715R_\odot$ (the base of the convection zone), the radiative diffusion transports all solar luminosity, i.e., $L_\odot$ (see Fig. 1.1h). This region is convectively stable. In the convection zone, the radiation is no longer efficient and the atmosphere becomes superadiabatically stratified, i.e., $ds/dr < 0$, where s is the specific entropy. Thus, this zone is characterized by turbulent thermal convection and the convective energy flux dominates over the radiative energy flux.

The convective energy flux, which is required to determine the temperature gradient and accordingly the solar structure, is estimated with the mixing length model (e.g. Stix 2004). In this model, the mixing length (l_{mix}), which is the mean free path for each convective parcel, is specified with a constant parameter α_{mix} as: $l_{\mathrm{mix}} = \alpha_{\mathrm{mix}}H_{\mathrm{p}}$, where H_{p} is the pressure scale height. It is assumed that each convective parcel is accelerated by buoyancy along the mixing length. Then a certain type of averaged vertical velocity and its convective energy flux can be estimated with the equation of motion. Once the convective energy flux is estimated one can solve equations to obtain the distribution of entropy and the mixing length parameter simultaneously (α_{mix}).

The results of the solar standard model are confirmed with global helioseismology data. The solar global oscillation observed at the photosphere is described with spherical harmonics (Y_{lm}) in horizontal space and Fourier transforms in time. The harmonic components are compared with the oscillations that are computed with the solar model after assuming linear and adiabatic perturbations. When the residuals of the model with respect to observations are assumed linear, inversion can be performed to obtain information on the solar interior (see also Stix 2004). Although there is an unexplained anomaly at the base of the convection zone, the residuals between the solar standard model and helioseismology are small. $\delta c_s^2 / c_s^2$ is typically 10^{-3} where c_s^2 is the square of the speed of sound and δc_s^2 is its difference (Basu et al. 1997).

1.1.2 Observation of Solar Mean Flow, Differential Rotation and Meridional Flow

Helioseismology has also revealed the mean structure of the solar flow such as differential rotation (global helioseismology) and meridional flow (local helioseismology).

The solar differential rotation on the surface was first observed by Christoph Scheiner as early as 1630 by using the motion of the sunspots. Then, in 1855 Carrington started the first quantitative observation of the solar rotation (see review by Beck 2000). Although several researchers expected the existence of a shear layer below the surface with different rotation rate between sunspots and Doppler velocity (see the introduction in Chap. 4), the internal structure of the solar differential rotation remained unknown until the advent of helioseismology. Solar rotation is fast enough to break the spherical symmetry of global oscillation and causes frequency splitting in terms of the azimuthal order m owing to the asymmetry in travel times between the eastward and westward waves (see review by Christensen-Dalsgaard 2002). Figure 1.2 shows the obtained result of the internal differential rotation estimated with data of HMI (data from Howe et al. 2011). The result shows five important features regarding the internal rotation. First the equator region rotates faster than the polar region. Note that the reliability of the data diminishes around the polar region. Second the distribution of the angular velocity is not cylindrical but conical in contrast to the Taylor-Proudman theory (see the theoretical discussion in Sect. 1.2). Third the radiation zone rotates almost rigidly at intermediate rate between the equator and the pole. Forth, there is a thin shear layer between the convection zone and radiation zone called tachocline. Although the details of the structure are controversial, the

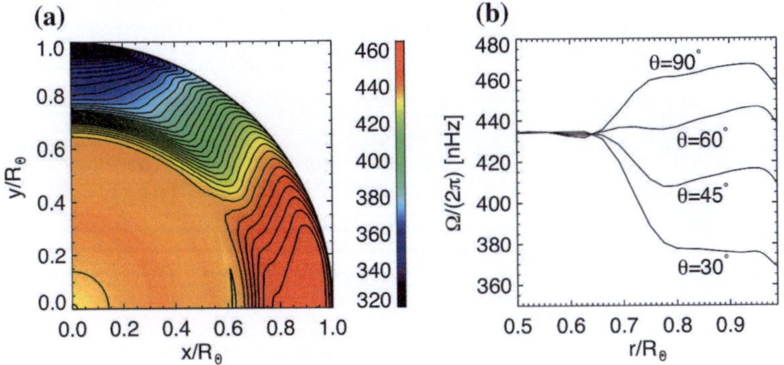

Fig. 1.2 An inversion of the helioseismology from HMI data about the angular velocity in the unit of nHz (Howe et al. 2011) **a** on meridional plane and **b** along the selected colatitude

tachocline is thought to be ellipsoidal. For example, Charbonneau et al. (1999) show that the center of the tachocline is $r_t/R_\odot = 0.693 \pm 0.003$ at the equator which is below the base of the convection zone and $r_t/R_\odot = 0.717 \pm 0.003$ at the pole which is slightly above the base. Although the thickness of the tachocline is still controversial, Charbonneau et al. (1999) show that the thickness is from $\Delta_t = 0.016R_\odot$ at the equator to $\Delta_t = 0.038R_\odot$ at the pole. The fifth important feature is the near surface shear layer (NSSL). There are significant deviations from the Taylor-Proudman state above $r = 0.95R_\odot$. In this layer, the angular velocity increases along the radius (see also the introduction of Chap. 4). The detailed distribution of the NSSL is studied by Corbard and Thompson (2002), using f modes from MDI data. They measured the gradient of the NSSL as about $-400\,\mathrm{nHz}/R_\odot$. The rotation rate was found to vary almost linearly with depth (Howe 2009). We note that the deviation from the Taylor-Proudman state in the NSSL is larger than that in the deep convection zone. This shear layer is the one of the targets of this thesis.

The meridional flow, known as the mean flow on the meridional plane, i.e., $\langle v_r \rangle$ and $\langle v_\theta \rangle$, is also observed on the surface with Doppler measurements (Hathaway et al. 1996; Hathaway 1996), where the parenthesis $\langle \rangle$ denotes the zonal average. These observations show the poleward meridional flow with the amplitude of a several $10\,\mathrm{m\,s^{-1}}$ (Giles et al. 1997). Using the global helioseismology, i.e., global-standing mode of the acoustic wave, it is difficult to distinguish the effect of the meridional flow as a perturbation of the standing-mode from those of magnetic field

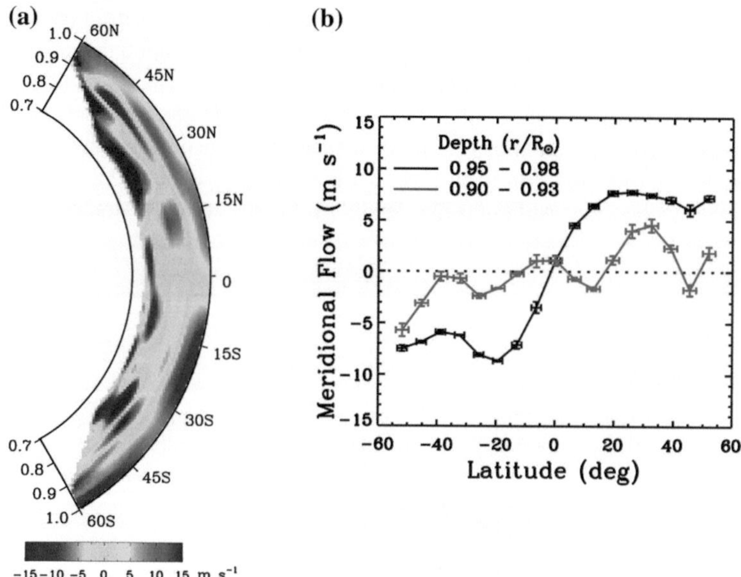

Fig. 1.3 The figure is from Zhao et al. (2013) with modifications. Meridional flow profile, obtained with the acoustic travel time. Panel **a** shows a flow profile on the meridional plane, with positive velocity directing northward. Panel **b** shows a flow profile as a function of latitude averaged over 0.90–0.93$R_\odot$ (*red line*) and 0.95–0.98$R_\odot$ (*black line*) (Color figure online)

and the centrifugal force since the amplitude of the meridional flow is relatively small ($\sim 10\,\mathrm{m\,s^{-1}}$) compared to the convection at the surface ($\sim 1\,\mathrm{km\,s^{-1}}$) and the speed of rotation ($2\,\mathrm{km\,s^{-1}}$). Duvall et al. (1993) suggested a new technique to investigate the flow structure in the solar interior; they called it "local helioseismology." In the time-distance method of local helioseismology, a correlation between two specific points is estimated. The travel time difference between the forward and backward waves is sensitive to the internal structure of the horizontal flow. Although the result show only a fraction of the meridional flow in the solar convection zone (the upper 4 % in radius and $\pm 60°$ in latitude), there is a poleward meridional flow in both hemispheres. Haber et al. (2002) also showed the asymmetry of the meridional flow about the equator and its dependence on time. In a certain phase of the solar cycle there is counter flow in the polar region.

Recent observation by Zhao et al. (2013) shows a 2D distribution of the meridional flow (Fig. 1.3a). Although the reliability in deeper layer is controversial, a return (equatorward) flow is seen in the middle of the convection zone and another cell exists in the lower part of the convection zone, i.e., multi-cell structure. In the near surface layer, the coherent poleward flow is observed, which is consistent with previous studies. Figure 1.3b indicates that the amplitude of the meridional flow increases along the radius. This result is important for understanding of the NSSL in Chap. 4.

1.2 Theory and Numerical Calculation for Differential Rotation and Meridional Flow

1.2.1 Gyroscopic Pumping and Thermal Wind Balance

To understand the maintenance mechanism of the mean flow, i.e., the differential rotation and meridional flow, the gyroscopic pumping and thermal wind balance equations, which we discuss in this section, are helpful (Rempel 2005; Miesch 2005; Miesch and Hindman 2011).

In the beginning, the Reynolds stress is reviewed. First, we consider the equation of motion in fluid dynamics:

$$\rho_0 \frac{\partial \mathbf{v}}{\partial t} = -\nabla \cdot (\rho_0 \mathbf{v}\mathbf{v}) - \nabla p, \tag{1.1}$$

where ρ_0, $\mathbf{v}$, and p is time-independent density, fluid velocity and gas pressure, respectively. Using a kind of ensemble average $\langle\rangle$, which is likely the zonal average at following discussion, the quantities (Q) are divided into the mean part $\langle Q \rangle$ and the perturbed part Q', i.e., $Q = \langle Q \rangle + Q'$. Then, the equation for the mean velocity $\langle \mathbf{v} \rangle$ is expressed as

$$\rho_0 \frac{\partial \langle \mathbf{v} \rangle}{\partial t} = -\nabla \cdot (\rho_0 \langle \mathbf{v} \rangle \langle \mathbf{v} \rangle) - \nabla \cdot (\rho_0 \langle \mathbf{v}'\mathbf{v}' \rangle) - \nabla \langle p \rangle, \tag{1.2}$$

where $\langle v' \rangle = 0$ is used. The terms $-\nabla \cdot (\rho_0 \langle \mathbf{v} \rangle \langle \mathbf{v} \rangle)$ and $-\nabla \langle p \rangle$ represent the processes from the mean quantities to the mean quantities. The term $-\nabla \cdot (\rho_0 \langle \mathbf{v}' \mathbf{v}' \rangle)$ represents the effect from the perturbed part to the mean part via the nonlinear coupling of the fluid velocity. The quantity $\rho_0 \langle \mathbf{v}' \mathbf{v}' \rangle$ is called the Reynolds stress. In anisotropic turbulence, nondiagonal terms in the Reynolds stress exist, which cause anisotropic momentum transport.

To derive the equations for gyroscopic pumping and thermal wind balance, we consider the equation of motion without the kinetic viscosity and the magnetic field contributions.

$$\rho_0 \frac{\partial \mathbf{u}}{\partial t} = -\nabla \cdot (\rho_0 \mathbf{u} \mathbf{u}) - \nabla p_1 - \rho_1 g \mathbf{e}_r, \tag{1.3}$$

where p_1, ρ_1, g, and $\mathbf{e}_r$ are the perturbed gas pressure, the perturbed density, the gravitational acceleration, and the unit vector along the radial direction. In the following discussion $\mathbf{u}$ and $\mathbf{v}$ are the fluid velocities at the inertial reference system and the rotating system, i.e., $\mathbf{u} = \mathbf{v} + r \sin \theta \Omega_0 \mathbf{e}_\phi$, where Ω_0 is the rotation rate of the system. The background stratification, ρ_0 and p_0, is assumed to be in spherically symmetric hydrostatic equilibrium, i.e.,

$$0 = -\frac{dp_0}{dr} - \rho_0 g. \tag{1.4}$$

Gyroscopic pumping is derived from the conservation equation for the angular momentum, i.e., originally from the zonal component of the equation of motion

$$\rho_0 \frac{\partial u_\phi}{\partial t} = -\frac{1}{r^2} \frac{\partial}{\partial r} \left(r^2 \rho_0 u_r u_\phi \right) - \frac{1}{r \sin \theta} \frac{\partial}{\partial \theta} \left(\sin \theta \rho_0 u_\theta u_\phi \right)$$
$$- \frac{1}{r \sin \theta} \frac{\partial}{\partial \phi} (\rho_0 u_\phi u_\phi) - \frac{\rho_0 u_\phi u_r}{r} - \frac{\cot \theta \rho_0 u_\phi u_\theta}{r}$$
$$- \frac{1}{r \sin \theta} \frac{\partial p_1}{\partial \phi}. \tag{1.5}$$

The zonal component means the ϕ-component in the spherical geometry (r, θ, ϕ), where r and θ are the radius and the colatitude, respectively. Then, we multiply $r \sin \theta$ and define the specific angular momentum as $\mathscr{L} = u_\phi r \sin \theta$. The equation becomes

$$\rho_0 \frac{\partial \mathscr{L}}{\partial t} = -\frac{1}{r} \frac{\partial}{\partial r} \left(r^2 \sin \theta \rho_0 u_r u_\phi \right) - \frac{\partial}{\partial \theta} \left(\sin \theta \rho_0 u_\theta u_\phi \right)$$
$$- \frac{\partial}{\partial \phi} (\rho_0 u_\phi u_\phi) - \rho_0 u_\phi u_r \sin \theta - \rho_0 u_\phi u_\theta \cos \theta - \frac{\partial p_1}{\partial \phi}$$
$$= -\frac{1}{r^2} \frac{\partial}{\partial r} \left[r^2 (r \sin \theta \rho_0 u_r u_\phi) \right] - \frac{1}{r \sin \theta} \frac{\partial}{\partial \theta} \left[\sin \theta (r \sin \theta \rho_0 u_\theta u_\phi) \right]$$
$$- \frac{1}{r \sin \theta} \frac{\partial}{\partial \phi} \left[r \sin \theta (\rho_0 u_\phi u_\phi + p_1) \right]$$
$$= -\nabla \cdot [r \sin \theta (\rho_0 \mathbf{u} u_\phi + p_1 \mathbf{e}_\phi)]. \tag{1.6}$$

Gyroscopic pumping shows the balance of the angular momentum transport on the meridional plane after the zonal average. When an anelastic approximation is valid for the mean flow ($\nabla \cdot (\rho_0 \langle \mathbf{v_m} \rangle) = 0$), Eq. (1.6) with zonal average is expressed as

$$\rho_0 \frac{\partial \langle \mathscr{L} \rangle}{\partial t} = -\rho_0 \langle \mathbf{v_m} \rangle \cdot \nabla \langle \mathscr{L} \rangle - \nabla \cdot (r \sin \theta \rho_0 \langle \mathbf{v'_m} v'_\phi \rangle). \tag{1.7}$$

Then the thermal wind balance equation is derived. The thermal wind balance equation is the originally zonal component of the vorticity equation. For a rotational system with rotation rate Ω_0, the equation of motion, after some algebra, is expressed using the Coriolis force as

$$\frac{\partial \mathbf{v}}{\partial t} = -(\mathbf{v} \cdot \nabla)\mathbf{v} - \frac{\nabla p_1}{\rho_0} - \frac{\rho_1}{\rho_0} g \mathbf{e}_r + 2\mathbf{v} \times \Omega_0, \tag{1.8}$$

where $\Omega_0 = \Omega_0 \mathbf{e_z}$ and $\mathbf{e_z} = \cos \theta \mathbf{e_r} - \sin \theta \mathbf{e}_\theta$ is the unit vector along the rotational axis (z-axis). Using the vector formula

$$(\mathbf{v} \cdot \nabla)\mathbf{v} = \nabla \left(\frac{v^2}{2} \right) - \mathbf{v} \times (\nabla \times \mathbf{v}), \tag{1.9}$$

and taking the curl of the equation of motion, the vorticity equation is obtained

$$\frac{\partial \omega}{\partial t} = \nabla \times (\mathbf{v} \times \omega) + \nabla \times \left(-\frac{\nabla p_1 + \rho_1 g \mathbf{e_r}}{\rho_0} \right) + \nabla \times (2\mathbf{v} \times \Omega_0). \tag{1.10}$$

In the thermal wind equation, the zonal component is focused. Then, the zonal component of the second and third terms in the right-hand side is computed as

$$\left[\nabla \times \left(-\frac{\nabla p_1 + \rho_1 g \mathbf{e_r}}{\rho_0} \right) \right]_\phi = \frac{1}{\rho_0^2 r} \frac{d\rho_0}{dr} \frac{\partial p_1}{\partial \theta} + \frac{g}{\rho_0 r} \frac{\partial \rho_1}{\partial \theta}$$

$$= -\frac{g}{\rho_0 r} \left[\left(\frac{\partial \rho}{\partial p} \right)_s \frac{\partial p_1}{\partial \theta} - \frac{\partial \rho_1}{\partial \theta} \right]$$

$$= \frac{g}{\rho_0 r} \left(\frac{\partial \rho}{\partial s} \right)_p \frac{\partial s_1}{\partial \theta}, \tag{1.11}$$

and

$$[\nabla \times (2\mathbf{v} \times \Omega_0)]_\phi = [2(\Omega_0 \cdot \nabla)\mathbf{v_r} - 2(\mathbf{v} \cdot \nabla)\Omega_0]_\phi$$

$$= 2(\Omega_0 \cdot \nabla)v_\phi = 2r \sin \theta \Omega_0 \frac{\partial \Omega_1}{\partial z}, \tag{1.12}$$

respectively. Subsequently, the zonal average for the vorticity equation is

$$\frac{\partial \langle \omega_\phi \rangle}{\partial t} = [\langle \nabla \times (\mathbf{v} \times \omega) \rangle]_\phi + 2r \sin \theta \, \Omega_0 \frac{\partial \langle \Omega_1 \rangle}{\partial z} + \frac{g}{\rho_0 r} \left(\frac{\partial \rho}{\partial s} \right)_p \frac{\partial \langle s_1 \rangle}{\partial \theta}. \quad (1.13)$$

The equations for the mean flows are thus derived. The equations are originally derived from the equation of motion, thus the gyroscopic pumping and the thermal wind balance equation show the time evolution of the angular momentum and the meridional flow, respectively. In the steady state ($\partial/\partial t = 0$), however, the gyroscopic pumping and the thermal wind balance equations determine the meridional flow and the differential rotation, respectively. In the steady state, gyroscopic pumping is

$$\rho_0 \langle \mathbf{v}_m \rangle \cdot \nabla \langle \mathscr{L} \rangle = -\nabla \cdot (r \sin \theta \rho_0 \langle \mathbf{v}'_m v'_\phi \rangle). \quad (1.14)$$

This equation indicates that when the Reynolds stress is given, the meridional flow $\langle \mathbf{v}_m \rangle$ can be determined. The thermal wind balance equation in the steady state is expressed as

$$- 2r \sin \theta \, \Omega_0 \frac{\partial \langle \Omega_1 \rangle}{\partial z} = [\langle \nabla \times (\mathbf{v} \times \omega) \rangle]_\phi + \frac{g}{\rho_0 r} \left(\frac{\partial \rho}{\partial s} \right)_p \frac{\partial \langle s_1 \rangle}{\partial \theta}. \quad (1.15)$$

This also indicates that when the advection/stretching term ($[\langle \nabla \times (\mathbf{v} \times \omega) \rangle]_\phi$) and the latitudinal entropy gradient are given, the differential rotation $\langle \Omega_1 \rangle$ is determined. Note that there is a possibility that the mean flows, $\langle \mathbf{v}_m \rangle$ and $\langle \Omega_1 \rangle$, affect the Reynolds stress in return.

The distribution of the Reynolds stress is required in the direct numerical calculations, otherwise models have to be used (Kitchatinov and Rüdiger 1995; Küker and Stix 2001; Rempel 2005; Hotta and Yokoyama 2011). The investigations that use direct numerical calculations are reviewed in the next section.

1.2.2 Numerical Calculations for Differential Rotation and Meridional Flow

There have been numerous studies about differential rotation using the mean-field model in which the thermal convection is treated as the parameterized effect (e.g. Kitchatinov and Rüdiger 1995; Küker and Stix 2001; Rempel 2005; Hotta and Yokoyama 2011) and the three-dimensional model in spherical shell including thermal convection using the anelastic approximation (Gilman and Miller 1981; Glatzmaier 1984; Miesch et al. 2000; Brun and Toomre 2002; Miesch et al. 2006, 2008; Brun et al. 2011). In this type of numerical calculations, the solar parameters in the standard model are adopted as the background stratification.

The latest conclusions from these studies are summarized as follows: (1) A banana cell-like convective structure causes the equatorward angular momentum transport (Miesch et al. 2000). (2) Radially inward angular momentum transport is established in almost the whole of the convection zone (Brun and Toomre 2002). (3) A latitudinal entropy gradient that corresponds to a temperature difference of $\sim$10 K between the pole and the equator is required to reproduce the conical profile (Miesch et al. 2006) and the tachocline (Brun et al. 2011).

The banana cell is especially established outside the tangential cylinder ($\sin\theta > r_{\mathrm{base}}/r$, where r_{base} is the location of the base of the convection zone). Because the Coriolis force is expressed as $2\rho_0\mathbf{v}\times\mathbf{\Omega}_0$, the fluid parcel rotates around the rotational axis when the disturbance from the boundary is not significant. This is analogous to the Larmor motion of the plasma particle, where the Lorentz force is proportional to $\mathbf{v}\times\mathbf{B}$. A topological explanation for the equatorward angular momentum transport is given in Miesch (2005). In the northern hemisphere, the prograde flow ($v'_\phi > 0$) is bent equatorward ($v'_\theta > 0$) and the retrograde flow ($v'_\phi < 0$) is bent poleward ($v'_\theta < 0$). As a result, the correlation is positive $\langle v'_\theta v'_\phi \rangle > 0$ (see Fig. 1.4). This implies equatorward angular momentum transport. More generally this is explained with the equation of motion on the perturbed velocity along the λ-direction (v'_λ), where λ is the direction perpendicular to the rotation axis. Note that the unit vector in λ is $\mathbf{e}_\lambda = \sin\theta\,\mathbf{e_r} + \cos\theta\,\mathbf{e}_\theta$. The equation of motion is expressed as

$$\frac{\partial v'_\lambda}{\partial t} = [\cdots] + 2v'_\phi\Omega_0. \qquad (1.16)$$

This shows that the banana cell is likely to generate positive correlation $\langle v'_\lambda v'_\phi \rangle > 0$, with outward angular momentum transport. This is the essential mechanism accelerating the equator.

Next we discuss the radial angular momentum transport. When the banana cell is not well established, the correlation between the radial and zonal velocities is generated from the radial component of the equation of motion.

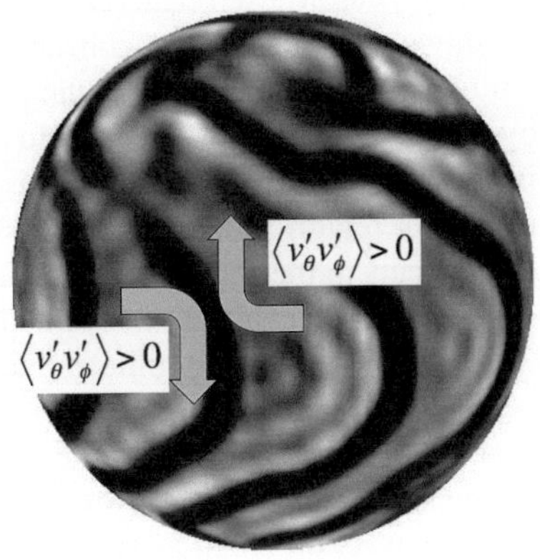

Fig. 1.4 Figure is from Miesch et al. (2000) with some modifications. At low latitude, there are banana cells

$$\frac{\partial v'_\phi}{\partial t} = [\cdots] - 2v'_r \Omega_0 \sin\theta. \tag{1.17}$$

Thus, a negative correlation $\langle v'_r v'_\phi \rangle < 0$ is generated. This implies downward angular momentum transport in the convection zone. In most of the calculations, the angular momentum flux by the Reynolds stress peaks in amplitude around the middle depth of the convection zone. This indicates that the value $-\nabla \cdot (\rho_0 r \sin\theta \langle v'_r v'_\phi \rangle \mathbf{e_r})$ is negative (positive) at the upper (lower) part of the convection zone. Even when the differential rotation is conical, the distribution of the angular momentum is almost cylindrical, i.e., $\nabla\langle \mathscr{L} \rangle \sim d\langle \mathscr{L} \rangle/d\lambda$, because of the factor of $(r^2 \sin^2\theta)$. Hence gyroscopic pumping becomes

$$\rho_0 \langle \mathbf{v_m} \rangle \frac{d\langle \mathscr{L} \rangle}{d\lambda} = -\nabla \cdot (\rho_0 r \sin\theta \langle v'_r v'_\phi \rangle \mathbf{e_r}). \tag{1.18}$$

Then poleward (equatorward) meridional flow is established at the upper (lower) part of the convection zone with a positive value of $d\langle \mathscr{L} \rangle/d\lambda$.

The discussion then moves to the thermal wind balance equation (1.15), i.e., the conical profile, the tachocline, and the NSSL. Equation (1.15) indicates that the contributions from the advection/stretching term and/or the entropy gradient are required to maintain the conical profile, the tachocline, and the NSSL, because they are in the non-Taylor-Proudman state ($\partial\langle \Omega_1 \rangle/\partial z \neq 0$) in steady state. In the convection zone, the entropy gradient is considered critical (the role of the advection/stretching term is discussed in Chap. 4). There are two possible mechanisms to generate the latitudinal entropy gradient. The first is an anisotropic correlation of the velocity and entropy $\langle v'_\theta s'_1 \rangle$. In the convection zone, the flow is likely aligned along the rotational axis and the correlation is negative $\langle v'_r v'_\theta \rangle < 0$. In thermal convection, the radial velocity and entropy fluctuation are well correlated $\langle v'_r s'_1 \rangle > 0$, because the hot (cool) plasma moves upward (downward). As a result, a negative correlation $\langle v'_\theta s'_1 \rangle$ that transports the positive entropy poleward is generated. Miesch et al. (2006) calculated that the temperature difference between the pole and the equator generated by this process is approximately 8 K which is not sufficiently large to explain the solar differential rotation. Thus, Miesch et al. (2006) added a latitudinal entropy gradient as a boundary condition at the base of the convection zone. The second mechanism to generate the latitudinal entropy gradient is the penetrating meridional flow originally suggested by Rempel (2005) in his mean-field model. When the anticlockwise meridional flow is established in the northern hemisphere, the downflow at the pole penetrates the overshoot region and generates positive entropy perturbation. The same phenomenon occurs in the upflow at the equator region and negative entropy perturbation is generated. Brun et al. (2011) simultaneously reproduced these two mechanisms and established the conical profile and tachocline in a self-consistent manner.

Finally in this section, the latest numerical calculations are introduced. Miesch et al. (2008) is a state-of-art study which considered the convection zone only. They achieved the highest resolution and provided the latest understanding of the physics in the convection zone. Their upper boundary is at $r = 0.98R_\odot$ and the resolution

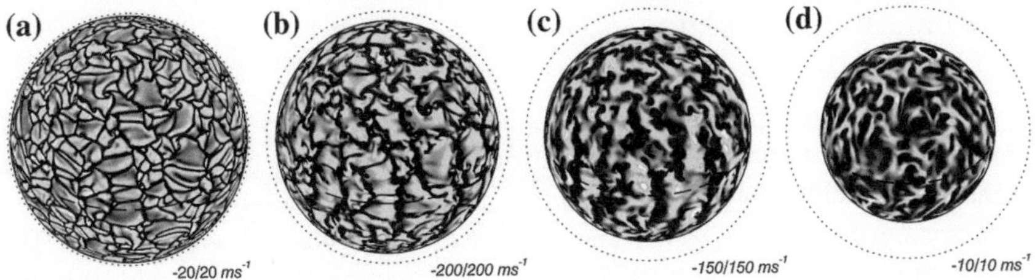

Fig. 1.5 Radial velocity v_r at selected depths in Miesch et al. (2008). **a** $0.98R_\odot$, **b** $0.92R_\odot$, **c** $0.85R_\odot$, and **d** $0.71R_\odot$

is $N_r \times N_\theta \times N_\phi = 257 \times 1{,}024 \times 2{,}048$, where N_r, N_θ, and N_ϕ are the number of grid points in the radial, latitudinal, and longitudinal direction, respectively. The horizontal grid spacing is approximately 2.2 Mm at the top boundary. Figure 1.5 shows the distribution of the radial velocity. The spectral peak of the radial velocity is estimated at $l \sim 80$, which corresponds to the horizontal scale of 55 Mm. On account of the low viscosity due to the high resolution, a relatively proper balance of the angular momentum transport between the meridional flow and the Reynolds stress is established. Figure 1.6 shows the results for the differential rotation, the meridional flow, and the temperature distribution. Miesch et al. (2008) added the latitudinal temperature gradient as boundary condition and established the solar-like conical profile of the differential rotation. The origin of the counter flows near the boundary was not well discussed; nonetheless the prominent one-cell meridional flow was reproduced in the convection zone.

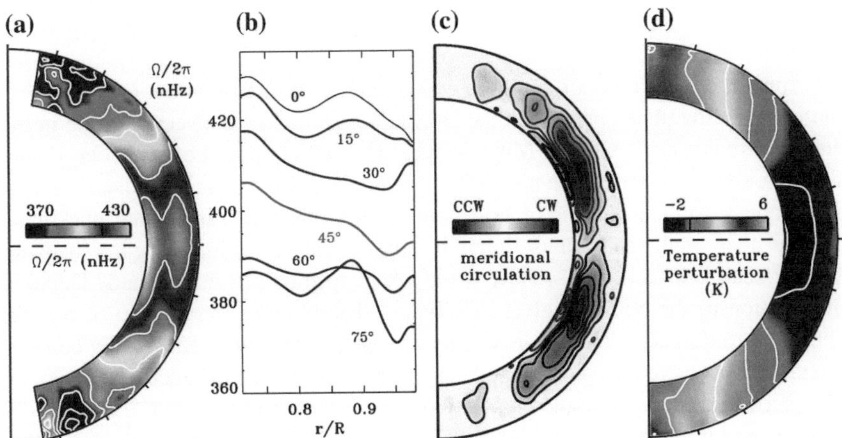

Fig. 1.6 **a** Differential rotation, **b** angular velocity at selected latitudes, **c** meridional flow, and **d** mean temperature perturbation in Miesch et al. (2008)

1.3 Remaining Problems

As discussed in the previous sections, observations, theory, and the numerical calculations have improved our understanding of the convection zone. There are essentially two remaining problems in the numerical calculation: (1) The difficulty in increasing the resolution. (2) The inaccessibility to the real solar surface. The causes are explained in the next section. In this section, the reason why they are required is explained.

There are several reasons why high resolution is required. Some are related to the magnetic field. Zwaan (1987) reported that the magnetic flux of sunspots is from 10^{20} to 10^{22} Mx. At the base of the convection zone, the magnetic strength is estimated to be 5×10^4 G to reproduce the tilt angle of the sunspot pair (Weber et al. 2011). Several studies suggest that magnetic field with this strength can be generated by explosion processes (Rempel and Schüssler 2001; Hotta et al. 2012a). These results indicate that the radius of the magnetic flux tube at the base of the convection zone is 2.5 Mm for the largest sunspot. This is comparable with the grid spacing of the current highest resolution calculation (Miesch et al. 2008). To avoid dissipation by numerical diffusivity, at least 10 grid points are required to resolve the flux tube. In addition, Cheung et al. (2006) suggested that the magnetic Reynolds number, which is determined by the resolution, has a significant impact on the behavior of the flux tube.

There is another requirement for the higher resolution owing to local (small scale) dynamo action. The dynamo, especially the stretching, is most effective in small scales. Although some numerical calculations in the local small box reveals the properties of the local dynamo on the photosphere (Vögler and Schüssler 2007; Pietarila Graham et al. 2010), the turbulent effect on the generation and the transportation of the magnetic field in the convection is unclear because it requires a huge number of grid points to resolve the inertial scale of the turbulence.

A fundamental and important issue which requires high resolution, is the connection between the photosphere and the convection zone. The convection scale in the photosphere is quite small ($\sim$1 Mm); hence, the calculation for both the solar global scale (the sun's circumference is 4,400 Mm) and the photosphere's small convection also requires a huge number of grid points. This raises two important questions. (1) How does small-scale convection in the near surface layer influence the structure of large-scale convection? (2) How is the NSSL formed and maintained? Detailed introduction to the NSSL is given in Chap. 4. The NSSL is thought to be a layer where the rotational influence drastically changes as the time scale of the convection changes. The requirement to address these problems are the accessibility to the solar surface as well as higher resolution. Anelastic approximation, the currently well-adopted method, however, has difficulties with both of them. A new method is adopted in this thesis.

1.4 Reduced Speed of Sound Technique

Before the problems with the anelastic approximation are explained, the reason why the anelastic approximation is adopted for the numerical calculations of the solar and stellar convection zone needs to be explained. One of the most significant difficulties arises from the large speed of sound, and the related low Mach-number flows throughout most of the convection zone. At the base of the convection zone, the speed of sound is approximately $200\,\text{km s}^{-1}$, whereas the speed of convection is thought to be $50\,\text{m s}^{-1}$ (e.g. Stix 2004). The time step must therefore be shorter on account of the CFL condition in an explicit fully compressible method even when we are interested in phenomena related to convection. To avoid this situation, the anelastic approximation is frequently adopted in which the mass conservation equation is replaced with $\nabla \cdot (\rho_0 \mathbf{v}) = 0$, where ρ_0 is the reference density and $\mathbf{v}$ is the fluid velocity. In this approximation, the speed of sound is assumed infinite and one needs to solve the elliptic equation for pressure, which filters out the propagation of the sound wave. Because the anelastic approximation is applicable deep in the convection zone and the time step is no longer limited by the high speed of sound, the solar global convection has been investigated with this method in many studies dealing with the differential rotation, the meridional flow, the global dynamo, and dynamical coupling of the radiative zone (Miesch et al. 2000, 2006, 2008; Brun and Toomre 2002; Brun et al. 2004, 2011; Browning et al. 2006; Ghizaru et al. 2010) as explained in the previous section.

There are, however, two drawbacks in the anelastic approximation. The first is the breakdown of the approximation near the solar surface. Because the convection velocity increases and the speed of sound decreases in the near surface layer ($>0.98 R_\odot$), they have similar values and the anelastic approximation cannot be applied. The connection between the near surface layer and the global convection is an ongoing challenge (e.g. Augustson et al. 2011). A global calculation, however, which includes all multiple scales, has not been achieved yet.

The second drawback is the difficulty in increasing the resolution. The pseudo-spectral method based on spherical harmonic expansion is frequently adopted, especially for solving the elliptic equation of pressure. In this method, the nonlinear terms require the transformation of physical variables from real space to spectral space and vice versa at every time step. The calculation cost of the transformation is estimated at $\mathcal{O}(N_\theta^2 N_\phi \log N_\phi)$ owing to the absence of a fast algorithm for the Legendre transformation, which is as powerful as the fast Fourier transformation (FFT), where N_θ and N_ϕ are the maximum mode numbers in latitude and longitude, respectively. Thus, the computational cost of this method is significant and limits the achievable resolution. Owing to this, several numerical calculations of the geodynamo adopt the finite difference method to achieve high resolution (Kageyama et al. 2008; Miyagoshi et al. 2010). As explained above, when the near surface layer is included in the calculations, the typical convection scale decreases and a large number of grid points is required. The resolution is critical to assess the solar surface. Note that several studies using the finite difference method have been performed in the stellar or solar context using a moderate ratio for the speed of sound and convection velocity by adjusting the radiative flux and stratification (Käpylä et al. 2011, 2012). Although this type of approach offers insight for the maintenance of the mean flow and the magnetic field, proper reproduction using solar parameters, such as stratification, luminosity, ionization effect, and rotation, and direct comparison with observations cannot be achieved.

The reduced speed of sound technique (here after RSST, Rempel 2005, 2006; Hotta et al. 2012b) can overcome such drawbacks while avoiding the severe time step caused by the speed of sound. In the RSST, the equation of continuity is replaced by

$$\frac{\partial \rho_1}{\partial t} = -\frac{1}{\xi^2} \nabla \cdot (\rho_0 \mathbf{v}), \tag{1.19}$$

where ρ_0 and ρ_1 are the background and perturbed density, respectively. Then the speed of sound is reduced ξ times, but the dispersion relation for sound waves remains; the wave speed decreases equally for all wavelengths. This technique does not change the hyperbolic character of the equations, which can be integrated explicitly. Owing to this hyperbolicity, only local communication is required. This decreases the communication overhead in parallel computing. Simple algorithms and low-cost communication significantly facilitate high-resolution calculations. Hotta et al. (2012b) investigated the validity of the RSST in a thermal convection problem. They concluded that the RSST is valid when the Mach number, defined using the root mean square (RMS) velocity and the reduced speed of sound $\hat{c}_s$, is smaller than 0.7. Another advantage of this method is the accessibility to the real solar surface with inhomogeneous ξ. The Mach number substantially varies in the solar convection zone. When moderate or no reduction in the speed of sound is used in the near surface laycr while using a large ξ around the bottom part of the convection zone, the properties of the thermal convection, even including the surface, is properly investigated without losing the physics. It is confirmed in Hotta et al. (2012b) that the inhomogeneous ξ is valid when the Mach number is less than 0.7.

1.5 Thesis Goals

This thesis has three goals.

1. To develop the numerical code for effectively managing the huge number of CPUs ($\sim 10^5$) in a good performance with the reduced speed of sound technique in the spherical geometry. Even with the RSST, there are further requirements for treating the near surface layer in spherical geometry, such as the partial ionization effect of hydrogen and helium and the severe time step caused by the convergence of the grid spacing around the pole. The development of such complex numerical code with good scaling and performance requires sophisticated algorithms and detailed tuning for specific supercomputers. These steps are shown in Chap. 2.

2. To achieve the unprecedented resolution and small-scale convection with unprecedented higher upper boundary and attain significant scale gap in the thermal convection between the middle of the convection zone and the near surface layer. This will enable us to address how small-scale convection in the near surface layer influences the convection in the deeper layers. In addition, higher resolution makes the convection significantly turbulent, which allows to better understand the generation and transport of small-scale magnetic field. These are discussed in Chap. 3. Note that, in Chap. 3, no rotation is taken into account to focus on the effects of turbulent convection.
3. To achieve the near surface shear layer with rotation. In this layer, the influence of the rotation is significantly different from that in the deeper convection zone. This means that the time and spatial scales of thermal convection change significantly. Although including these scales in the numerical calculations is difficult and challenging, the proposed high-performance method and numerical code can do it. This is discussed in Chap. 4.

In Chap. 5, we summarize the thesis results and discuss the conclusion.

References

K. Augustson, M. Rast, R. Trampedach, J. Toomre, Modeling the near-surface shear layer: diffusion schemes studied with CSS. J. Phys. Conf. Ser. **271**(1), 012070 (2011). doi:10.1088/1742-6596/271/1/012070

S. Basu, J. Christensen-Dalsgaard, W.J. Chaplin, Y. Elsworth, G.R. Isaak, R. New, J. Schou, M.J. Thompson, S. Tomczyk, Solar internal sound speed as inferred from combined BiSON and LOWL oscillation frequencies. MNRAS **292**, 243 (1997)

J.G. Beck, A comparison of differential rotation measurements—(Invited Review). Sol. Phys. **191**, 47–70 (2000)

M.K. Browning, M.S. Miesch, A.S. Brun, J. Toomre, Dynamo action in the solar convection zone and tachocline: pumping and organization of toroidal fields. ApJ **648**, 157–160 (2006). doi:10.1086/507869

A.S. Brun, J. Toomre, Turbulent convection under the influence of rotation: sustaining a strong differential rotation. ApJ **570**, 865–885 (2002). doi:10.1086/339228

A.S. Brun, M.S. Miesch, J. Toomre, Global-scale turbulent convection and magnetic dynamo action in the solar envelope. ApJ **614**, 1073–1098 (2004). doi:10.1086/423835

A.S. Brun, M.S. Miesch, J. Toomre, Modeling the dynamical coupling of solar convection with the radiative interior. ApJ **742**, 79 (2011). doi:10.1088/0004-637X/742/2/79

P. Charbonneau, J. Christensen-Dalsgaard, R. Henning, R.M. Larsen, J. Schou, M.J. Thompson, S. Tomczyk, Helioseismic constraints on the structure of the solar tachocline. ApJ **527**, 445–460 (1999). doi:10.1086/308050

M.C.M. Cheung, F. Moreno-Insertis, M. Schüssler, Moving magnetic tubes: fragmentation, vortex streets and the limit of the approximation of thin flux tubes. A&A **451**, 303–317 (2006). doi:10.1051/0004-6361:20054499

J. Christensen-Dalsgaard, Helioseismology. Rev. Mod. Phys. **74**, 1073–1129 (2002). doi:10.1103/RevModPhys.74.1073

J. Christensen-Dalsgaard, W. Dappen, S.V. Ajukov, E.R. Anderson, H.M. Antia, S. Basu, V.A. Baturin, G. Berthomieu, B. Chaboyer, S.M. Chitre, A.N. Cox, P. Demarque, J. Donatowicz, W.A. Dziembowski, M. Gabriel, D.O. Gough, D.B. Guenther, J.A. Guzik, J.W. Harvey, F. Hill, G.

Houdek, C.A. Iglesias, A.G. Kosovichev, J.W. Leibacher, P. Morel, C.R. Proffitt, J. Provost, J. Reiter, E.J. Rhodes Jr, F.J. Rogers, I.W. Roxburgh, M.J. Thompson, R.K. Ulrich, The current state of solar modeling. Science **272**, 1286–1292 (1996). doi:10.1126/science.272.5266.1286

T. Corbard, M.J. Thompson, The subsurface radial gradient of solar angular velocity from MDI f-mode observations. Sol. Phys. **205**, 211–229 (2002). doi:10.1023/A:1014224523374

T.L. Duvall Jr., S.M. Jefferies, J.W. Harvey, M.A. Pomerantz, Time-distance helioseismology. Nature **362**, 430–432 (1993). doi:10.1038/362430a0

M. Ghizaru, P. Charbonneau, P.K. Smolarkiewicz, Magnetic cycles in global large-eddy simulations of solar convection. ApJL **715**, 133–137 (2010). doi:10.1088/2041-8205/715/2/L133

P.M. Giles, T.L. Duvall, P.H. Scherrer, R.S. Bogart, A subsurface flow of material from the Sun's equator to its poles. Nature **390**, 52–54 (1997). doi:10.1038/36294

P.A. Gilman, J. Miller, Dynamically consistent nonlinear dynamos driven by convection in a rotating spherical shell. ApJS **46**, 211–238 (1981). doi:10.1086/190743

G.A. Glatzmaier, Numerical simulations of stellar convective dynamos. I—the model and method. J. Comput. Phys. **55**, 461–484 (1984). doi:10.1016/0021-9991(84)90033-0

D.A. Haber, B.W. Hindman, J. Toomre, R.S. Bogart, R.M. Larsen, F. Hill, Evolving submerged meridional circulation cells within the upper convection zone revealed by ring-diagram analysis. ApJ **570**, 855–864 (2002). doi:10.1086/339631

D.H. Hathaway, Doppler measurements of the Sun's meridional flow. ApJ **460**, 1027 (1996). doi:10. 1086/177029

D.H. Hathaway, P.A. Gilman, J.W. Harvey, F. Hill, R.F. Howard, H.P. Jones, J.C. Kasher, J.W. Leibacher, J.A. Pintar, G.W. Simon, GONG observations of solar surface flows. Science **272**, 1306–1309 (1996). doi:10.1126/science.272.5266.1306

H. Hotta, T. Yokoyama, Modeling of differential rotation in rapidly rotating solar-type stars. ApJ **740**, 12 (2011). doi:10.1088/0004-637X/740/1/12

H. Hotta, M. Rempel, T. Yokoyama, Magnetic field intensification by the three-dimensional "Explosion" process. ApJL **759**, 24 (2012a). doi:10.1088/2041-8205/759/1/L24

H. Hotta, M. Rempel, T. Yokoyama, Y. Iida, Y. Fan, Numerical calculation of convection with reduced speed of sound technique. A&A **539**, 30 (2012b). doi:10.1051/0004-6361/201118268

R. Howe, Solar interior rotation and its variation. Living Rev. Sol. Phys. **6**, 1 (2009). doi:10.12942/ lrsp-2009-1

R. Howe, T.P. Larson, J. Schou, F. Hill, R. Komm, J. Christensen-Dalsgaard, M.J. Thompson, First global rotation inversions of HMI data. J. Phys. Conf. Ser. **271**(1), 012061 (2011). doi:10.1088/ 1742-6596/271/1/012061

A. Kageyama, T. Miyagoshi, T. Sato, Formation of current coils in geodynamo simulations. Nature **454**, 1106–1109 (2008). doi:10.1038/nature07227

P.J. Käpylä, M.J. Mantere, A. Brandenburg, Cyclic magnetic activity due to turbulent convection in spherical wedge geometry. ApJL **755**, 22 (2012). doi:10.1088/2041-8205/755/1/L22

P.J. Käpylä, M.J. Mantere, G. Guerrero, A. Brandenburg, P. Chatterjee, Reynolds stress and heat flux in spherical shell convection. A&A **531**, 162 (2011). doi:10.1051/0004-6361/201015884

L.L. Kitchatinov, G. Rüdiger, Differential rotation in solar-type stars: revisiting the Taylor-number puzzle. A&A **299**, 446 (1995)

M. Küker, M. Stix, Differential rotation of the present and the pre-main-sequence Sun. A&A **366**, 668–675 (2001). doi:10.1051/0004-6361:20010009

M.S. Miesch, Large-scale dynamics of the convection zone and tachocline. Living Rev. Sol. Phys. **2**, 1 (2005)

M.S. Miesch, B.W. Hindman, Gyroscopic pumping in the solar near-surface shear layer. ApJ **743**, 79 (2011). doi:10.1088/0004-637X/743/1/79

M.S. Miesch, A.S. Brun, J. Toomre, Solar differential rotation influenced by latitudinal entropy variations in the tachocline. ApJ **641**, 618–625 (2006). doi:10.1086/499621

M.S. Miesch, A.S. Brun, M.L. De Rosa, J. Toomre, Structure and evolution of giant cells in global models of solar convection. ApJ **673**, 557–575 (2008). doi:10.1086/523838

M.S. Miesch, J.R. Elliott, J. Toomre, T.L. Clune, G.A. Glatzmaier, P.A. Gilman, Three-dimensional spherical simulations of solar convection. I. Differential rotation and pattern evolution achieved with laminar and turbulent states. ApJ **532**, 593–615 (2000). doi:10.1086/308555

T. Miyagoshi, A. Kageyama, T. Sato, Zonal flow formation in the Earth's core. Nature **463**, 793–796 (2010). doi:10.1038/nature08754

J. Pietarila Graham, R. Cameron, M. Schüssler, Turbulent small-scale dynamo action in solar surface simulations. ApJ **714**(2), 1606 (2010). http://stacks.iop.org/0004-637X/714/i=2/a=1606

M. Rempel, Solar differential rotation and meridional flow: the role of a subadiabatic tachocline for the Taylor-Proudman balance. ApJ **622**, 1320–1332 (2005). doi:10.1086/428282

M. Rempel, Flux-transport dynamos with Lorentz force feedback on differential rotation and meridional flow: saturation mechanism and torsional oscillations. ApJ **647**, 662–675 (2006). doi:10.1086/505170

M. Rempel, M. Schüssler, Intensification of magnetic fields by conversion of potential energy. ApJL **552**, 171–174 (2001). doi:10.1086/320346

M. Stix, *The Sun: An Introduction*, 2nd edn. Astronomy and Astrophysics Library (Springer, Berlin, 2004). ISBN: 3-540-20741-4

A. Vögler, M. Schüssler, A solar surface dynamo. A&A **465**, 43–46 (2007). doi:10.1051/0004-6361:20077253

M.A. Weber, Y. Fan, M.S. Miesch, The rise of active region flux tubes in the turbulent solar convective envelope. ApJ **741**, 11 (2011). doi:10.1088/0004-637X/741/1/11

J. Zhao, R.S. Bogart, A.G. Kosovichev, T.L. Duvall Jr, T. Hartlep, Detection of equatorward meridional flow and evidence of double-cell meridional circulation inside the sun. ApJL **774**, 29 (2013). doi:10.1088/2041-8205/774/2/L29

C. Zwaan, Elements and patterns in the solar magnetic field. ARA&A **25**, 83–111 (1987). doi:10.1146/annurev.aa.25.090187.000503

Chapter 2
Basic Equations and Development of Numerical Code

Abstract In this chapter, we show our development of numerical code. The detailed setting for calculations in the solar convection zone is introduced. The near-surface region ($>0.98R_\odot$) is included for the solar global convection calculation for the first time. Our new challenge for including the partial ionization effect of Hydrogen and Helium is explained. In order to deal with large number of CPUs in the huge parallel computer, we adopt efficient parallelizaion method, Peano-Hilbert space-filling curve and Yin-Yang grid. These are explained. Finally the method for analyzing the huge size data outputted by the calculations are introduced.

Keywords Yin-Yang grid · Artificial viscosity · Managing huge data · Equation of state including partial ionization

2.1 Model Setting

2.1.1 Equations

We solve the three-dimensional magnetohydrodynamic equations with the RSST in the spherical geometry (r, θ, ϕ) as follows:

$$\frac{\partial \rho_1}{\partial t} = -\frac{1}{\xi^2} \nabla \cdot (\rho_0 \mathbf{v}), \tag{2.1}$$

$$\rho_0 \frac{\partial \mathbf{v}}{\partial t} = -\rho_0 (\mathbf{v} \cdot \nabla)\mathbf{v} - \nabla \left(p_1 + \frac{B^2}{8\pi} \right) + \nabla \cdot \left(\frac{\mathbf{BB}}{4\pi} \right)$$
$$- \rho_1 g \mathbf{e_r} + 2\rho_0 \mathbf{v} \times \mathbf{\Omega}_0, \tag{2.2}$$

$$\frac{\partial \mathbf{B}}{\partial t} = \nabla \times (\mathbf{v} \times \mathbf{B}), \tag{2.3}$$

$$\rho_0 T_0 \frac{\partial s_1}{\partial t} = -\rho_0 T_0 (\mathbf{v} \cdot \nabla)s_1 + \frac{1}{r^2} \frac{d}{dr}\left(r^2 \kappa_{\mathrm{r}} \rho_0 c_{\mathrm{p}} \frac{dT_0}{dr} \right) + \Gamma, \tag{2.4}$$

where $\rho, p, s, T, \mathbf{v}$, and $\mathbf{B}$ are the density, gas pressure, specific entropy, temperature, fluid velocity, and magnetic field, respectively. Subscript 1 denotes the fluctuation

© Springer Japan 2015

H. Hotta, *Thermal Convection, Magnetic Field, and Differential Rotation in Solar-type Stars*, Springer Theses, DOI 10.1007/978-4-431-55399-1_2

"

from the time-independent spherically symmetric reference state, which has subscript 0. g, κ_r, and Γ are the gravitational acceleration, the coefficient of the radiative diffusivity, and the surface cooling term, respectively. Ω_0 is the rotating vector of the rotating frame. The setting of these values is explained in the following paragraph. The equation of state required to close the MHD system is explained in Sect. 2.1.5.

Note that we do not have any explicit turbulent thermal diffusivity and viscosity (Miesch et al. 2000) to maximize the fluid and magnetic Reynolds number but use the artificial viscosity introduced in Rempel et al. (2009). This method is explained in Sect. 2.2.2.

2.1.2 Background Stratification and Radiation

Figure 2.1 shows the reference state used in this study in comparison with Model S (Christensen-Dalsgaard et al. 1996). The reference stratification is determined by solving the one-dimensional hydrostatic equation and the realistic equation of state

$$\frac{dp_0}{dr} = -\rho_0 g, \tag{2.5}$$

$$\rho_0 = \rho_0(p_0, s_0), \tag{2.6}$$

$$\frac{ds_0}{dr} = 0. \tag{2.7}$$

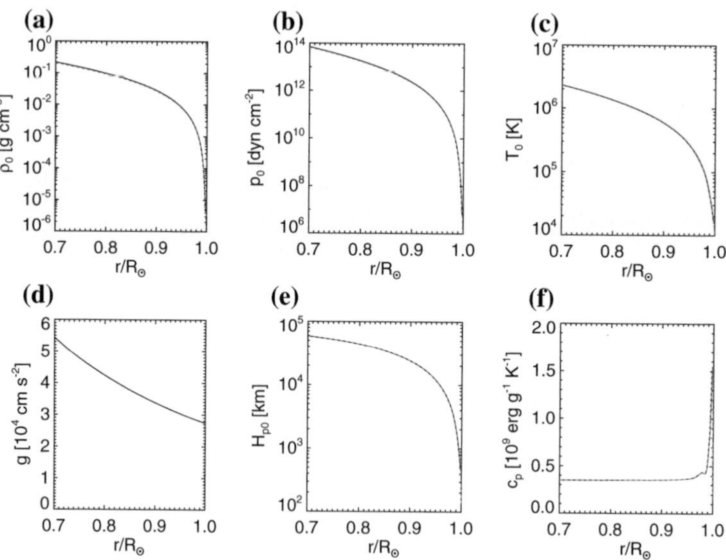

Fig. 2.1 The values at the reference state for **a** density, **b** gas pressure, **c** temperature, **d** gravitational acceleration, **e** pressure scale height, **f** heat capacity at constant pressure. The *black* and *red lines* show the reference state in this study and the values from Model S, respectively. The value of the gravitational acceleration in this study is exactly the same as Model S (Color figure online)

Equation (2.6) is calculated with the OPAL repository including partial ionization. The adiabatic stratification is set as the reference state and initial state. The stratification becomes superadiabatic after the development of convection as a consequence of radiative heating near the bottom and radiative cooling at the top as described below. The gravitational acceleration and the radiative diffusion are adopted from Model S. The boundary is set at $r = 0.998R_\odot$ with the values from Model S and the equations are integrated inward with the fourth-order Runge-Kutta method. A total of 3,072 grid points are used in the integration and stored. For each of the time-dependent calculations in this study, the interpolated values to each grid (i.e., variables with subscript 0 in Eqs. (2.1) and (2.4)) are from the stored data.

In the real sun, the surface is continuously cooled by radiation. Because the adopted boundary is not located at the real solar surface even though it is unprecedentedly closer to the real surface, we add artificial cooling (Γ) in Eq. (2.4)

$$\Gamma(r) = -\frac{1}{r^2}\frac{\partial}{\partial r}(r^2 F_s), \tag{2.8}$$

$$r^2 F_s(r) = r_{\min}^2 F_r(r_{\min})\exp\left[-\left(\frac{r - r_{\max}}{d_c}\right)^2\right], \tag{2.9}$$

$$F_r(r) = -\kappa_r \rho_0 c_p \frac{dT_0}{dr}, \tag{2.10}$$

where $r_{\min}$ and $r_{\max}$ denote the location of the bottom boundary and top boundary, respectively. This procedure ensures that the radiative luminosity inputted from the bottom is released through the top boundary. The realistic simulation for the near surface layer shows that the thickness of the cooling layer by radiation is similar to the local pressure scale height (e.g. Stein et al. 2009). Typically, two pressure scale heights are adopted for the thickness of the cooling layer, i.e., $d_c = 2H_{p0}(r_{\max})$, where $H_{p0} = p_0/(\rho_0 g)$ is the pressure scale height.

2.1.3 Setting for RSST

In this thesis, the factor of the RSST is set to make the adiabatic reduced speed of sound uniform in space. The adiabatic speed of sound is defined as:

$$c_s(r) = \sqrt{\left(\frac{\partial p}{\partial \rho}\right)_s}. \tag{2.11}$$

Then the factor of the RSST is set as

$$\xi(r) = \xi_0 \frac{c_s(r)}{c_s(r_{\min})}. \tag{2.12}$$

In this thesis, we adopt $\xi_0 = 120$ for calculations in Chap. 3. Thus, the reduced speed of sound $\hat{c}_s \equiv c_s/\xi = 1.88$ km s^{-1} at all depths. Hotta et al. (2012) suggested that the RSST is valid for thermal convection under the criterion of $v_{rms}/\hat{c}_s < 0.7$, where v_{rms} is the RMS (root mean square) convection velocity. Thus, the we can properly treat the convection with $v_{rms} < 1.3$ km s^{-1} in this model. In Chap. 4, since $\xi = 200$ is adopted, the valid convection speed is less than 0.79 km s^{-1}. Hotta et al. (2012) suggest that the total mass is not conserved with inhomogeneous ξ and long-term drift is not avoided from the reference state, i.e., mass continuously decreases or increases. In this study, however, we adopt different way to avoid this type of long-term drift. When the equation of continuity is treated as

$$\frac{\partial}{\partial t}\left(\xi^2 \rho_1\right) = -\nabla \cdot (\rho_0 \mathbf{v}),\tag{2.13}$$

the value $\hat{M}$ is conserved in the rounding error with appropriate boundary conditions, in which

$$\hat{M} = \int_V \xi^2 \rho_1 dV.\tag{2.14}$$

Although the radial distribution of the density is different from the original, the fluctuation remains small (e.g., $\rho_1/\rho_0 \sim 10^{-6}$) and does not affect the character of the thermal convection. Hotta et al. (2012) confirmed that the statistical features are not influenced by the inhomogeneous ξ.

2.1.4 Divergence Free Condition for Magnetic Field

The divergence free condition, i.e. $\nabla \cdot \mathbf{B} = 0$, is maintained with the diffusion scheme for each Runge-Kutta loop. This was also introduced by Rempel et al. (2009). The original idea for this is that when the diffusion equation for the divergence of magnetic field with an appropriate boundary condition, i.e., $\nabla \cdot \mathbf{B} = 0$, the numerical generated divergence error disappears. Thus the following equation is intended to be solved

$$\frac{\partial}{\partial t}(\nabla \cdot \mathbf{B}) = \nabla \cdot [\mu \nabla(\nabla \cdot \mathbf{B})],\tag{2.15}$$

where μ is the diffusion coefficient and adopted as large as possible for the CFL condition. In the numerical calculation, the equation:

$$\frac{\partial \mathbf{B}}{\partial t} = \mu \nabla(\nabla \cdot \mathbf{B}),\tag{2.16}$$

is solved and is mathematically identical with Eq. (2.15).

2.1.5 Equation of State

Since our upper boundary is at $r = 0.99R_\odot$ at maximum, in the near-surface region partial ionization is important (see Fig. 2.1f) and is included in our treatment by using the OPAL repository with solar abundances of $X = 0.75$, $Y = 0.23$, and $Z = 0.02$, where X, Y, and Z specify the mass fraction of hydrogen, helium and other heavy elements, respectively.

In the numerical simulations of the convection in the near surface layer, the ordinary tabular equation of state is widely used (Vögler et al. 2005; Rempel et al. 2009). However, this is not a good approach in our current simulations, because the deviations from the reference state are small, e.g. $\rho_1/\rho_0 \sim 10^{-6}$ around the base of the convection zone. Thus we adopt another way to treat the ionization effect in the near surface layer. The fluctuations from the reference state are calculated as

$$p_1 = \left(\frac{\partial p}{\partial \rho}\right)_s \rho_1 + \left(\frac{\partial p}{\partial s}\right)_\rho s_1, \tag{2.17}$$

$$T_1 = \left(\frac{\partial T}{\partial \rho}\right)_p \rho_1 + \left(\frac{\partial T}{\partial p}\right)_\rho p_1, \tag{2.18}$$

$$e_1 = \left(\frac{\partial e}{\partial \rho}\right)_T \rho_1 + \left(\frac{\partial e}{\partial T}\right)_\rho T_1, \tag{2.19}$$

where e is the internal energy. The first derivatives, such as $(\partial p/\partial \rho)_s$, are described by the background variables, $\rho_0(r)$, $p_0(r)$ ... and are regarded as functions of depth r. In the OPAL routine (Rogers et al. 1996), the values $(\partial e/\partial \rho)_T$, $(\partial e/\partial T)_\rho$, $(\partial p/\log \rho)_T$ and $(\partial p/\log T)_\rho$ are provided for given ρ_0, T_0, and the mass fractions of hydrogen (X), helium (Y) and other heavy elements (Z). The relations between the OPAL-provided variables and the required variables derived from the first law of thermodynamics are (Mihalas and Mihalas 1984):

$$\left(\frac{\partial p}{\partial \rho}\right)_s = \frac{c_p}{\kappa_t \rho_0 c_v}, \tag{2.20}$$

$$\left(\frac{\partial p}{\partial s}\right)_\rho = \frac{\beta_p T_0}{\kappa_t c_v}, \tag{2.21}$$

$$\left(\frac{\partial T}{\partial \rho}\right)_p = -\frac{1}{\rho_0 \beta_p}, \tag{2.22}$$

$$\left(\frac{\partial T}{\partial p}\right)_\rho = T_0 \bigg/ \left(\frac{\partial p}{\partial \log T}\right)_\rho, \tag{2.23}$$

where β_p, c_v, c_p and κ_t are the coefficient of thermal expansion, the specific heat at constant volume and pressure, and the coefficient of isothermal compressibility, respectively and they are defined as follows:

$$\beta_{\mathrm{p}} = -\left(\frac{\partial \log \rho}{\partial T}\right)_p = \frac{1}{T_0}\left(\frac{\partial p}{\partial \log T}\right)_\rho \Big/ \left(\frac{\partial p}{\partial \log \rho}\right)_T, \tag{2.24}$$

$$c_{\mathrm{v}} = \left(\frac{\partial e}{\partial T}\right)_\rho, \tag{2.25}$$

$$c_{\mathrm{p}} = c_{\mathrm{v}} - T_0\beta_{\mathrm{p}}\left[\left(\frac{\partial e}{\partial \rho}\right)_T - \left(\frac{p_0}{\rho_0^2}\right)\right], \tag{2.26}$$

$$\kappa_{\mathrm{t}} = \left(\frac{\partial \log \rho}{\partial p}\right)_T. \tag{2.27}$$

2.2 Numerical Method

2.2.1 Space Derivative and Time Integration

The numerical method used in this thesis is the same as that in the MURaM code (Vögler et al. 2005). We use the fourth-order space-centered difference for each derivative. The first spatial derivatives of quantity u are given by

$$\left(\frac{\partial u}{\partial x}\right)_i = \frac{1}{12\Delta x}\left(-u_{i+2} + 8u_{i+1} - 8u_{i-1} + u_{i-2}\right), \tag{2.28}$$

where i denotes the index of the grid position along a particular spatial direction. The numerical solution of the system is advanced in time with an explicit fourth-order Runge-Kutta scheme. The system of partial equations is written as

$$\frac{\partial \mathbf{U}}{\partial t} = \mathbf{R}(\mathbf{U}), \tag{2.29}$$

for $\mathbf{U}_{n+1}$, which is the value at $t_{n+1} = (n+1)\Delta t$. This is calculated in four steps as

$$\mathbf{U}_{n+\frac{1}{4}} = \mathbf{U}_n + \frac{\Delta t}{4}\mathbf{R}\left(\mathbf{U}_n\right), \tag{2.30}$$

$$\mathbf{U}_{n+\frac{1}{3}} = \mathbf{U}_n + \frac{\Delta t}{3}\mathbf{R}\left(\mathbf{U}_{n+\frac{1}{4}}\right), \tag{2.31}$$

$$\mathbf{U}_{n+\frac{1}{2}} = \mathbf{U}_n + \frac{\Delta t}{2}\mathbf{R}\left(\mathbf{U}_{n+\frac{1}{3}}\right), \tag{2.32}$$

$$\mathbf{U}_{n+1} = \mathbf{U}_n + \Delta t\mathbf{R}\left(\mathbf{U}_{n+\frac{1}{2}}\right). \tag{2.33}$$

2.2.2 Artificial Viscosity

The artificial viscosity used in this thesis is introduced by Rempel et al. (2009). The defined diffusive flux is

$$F_{i+1/2} = -\frac{1}{2}c_{i+1/2}\phi_{i+1/2}\left(u_{\mathrm{r}} - u_{\mathrm{l}}, u_{i+1} - u_i\right)\left(u_{\mathrm{r}} - u_{\mathrm{l}}\right), \tag{2.34}$$

$$\phi = \begin{cases} \left[\dfrac{(u_{\mathrm{r}} - u_{\mathrm{l}})}{(u_{i+1} - u_i)}\right]^2 & \text{for } (u_{\mathrm{r}} - u_{\mathrm{l}}) \cdot (u_{i+1} - u_i) > 0, \\ 0 & \text{for } (u_{\mathrm{r}} - u_{\mathrm{l}}) \cdot (u_{i+1} - u_i) \le 0, \end{cases} \tag{2.35}$$

where $c_{i+1/2} = c_{\mathrm{s}} + v + c_{\mathrm{a}}$ is the characteristic velocity which is the sum of the speed of sound (c_{s}), fluid velocity (v) and the Alfven velocity (c_{a}). To decrease the effect of viscosity, a multiplier less than unity is sometimes used. In the code, the physical variables u_i are defined at the center of the cell. To calculate the diffusive flux, the variables u_{r} and u_{l} at a boundary of the cells are defined as:

$$u_{\mathrm{l}} = u_i + \frac{1}{2}\Delta u_i, \tag{2.36}$$

$$u_{\mathrm{r}} = u_{i+1} - \frac{1}{2}\Delta u_{i+1}, \tag{2.37}$$

where the tilt of the variable (Δu_i) is defined as:

$$\Delta u_i = \text{minimod}\left(\varepsilon(u_{i+1} - u_i), \frac{u_{i+1} - u_{i-1}}{2}, \varepsilon(u_i - u_{i-1})\right), \tag{2.38}$$

where ε is the factor for the minimod function ($1 < \varepsilon < 2$).

To conserve total energy, the heat from the dissipated kinetic energy and the magnetic energy should be properly treated. The treatment for ensuring that the dissipated kinetic and magnetic energies with artificial viscosity are converted to internal energy is the following. The equation of motion and the induction equation are expressed as

$$\rho_0 \frac{\partial \mathbf{v}}{\partial t} = [\cdots] - \nabla \cdot [\rho_0 \mathbf{F}_{\mathrm{diff}}(\mathbf{v})],$$

$$\frac{\partial \mathbf{B}}{\partial t} = [\cdots] - \nabla \cdot [\mathbf{F}_{\mathrm{diff}}(\mathbf{B})], \tag{2.39}$$

where $\mathbf{F}_{\mathrm{diff}}$ is the diffusive flux calculated with Eq. (2.35). The heat caused by the artificial viscosity is estimated and added in the equation of entropy as

$$\rho T \frac{Ds}{Dt} = -[\rho_0 \mathbf{F}_{\mathrm{diff}}(\mathbf{v}) \cdot \nabla] \cdot \mathbf{v} - \frac{1}{4\pi}[\mathbf{F}_{\mathrm{diff}}(\mathbf{B}) \cdot \nabla] \cdot \mathbf{v}. \tag{2.40}$$

2.2.3 Peano-Hilbert Space Filling Curve for MPI Communication

For parallel computation by using the MPI library, the data for the spatial elements, each of which corresponds to an MPI process, should be loaded to reduce the communication among processes. For example, 64 MPI processes are used to divide the calculation space in two dimensions as $S_x \times S_y = 8 \times 8$, where S_x, and S_y are the number of data elements for the x-, and y-direction, respectively. As is often the case, the communication between neighbors is more effective than that between apart two, the communication in the x-direction is relatively effective, because the neighborhood in the x-direction is also a neighborhood in the MPI process number. Regarding the y-direction, it is not the case. In this case, the MPI processes are first assigned in the x-direction, and after one line is filled in the x-direction, they move to the y-direction. The neighborhood in the y-direction is always apart 8 in MPI process numbers. When the computer system is one-dimensional, this assignment decreases the efficiency of the communication cost. When the MPI process number becomes more than 10^3, this becomes even more problematic.

To avoid this problem, the Peano-Hilbert space-filling curve is adopted for ordering the MPI rank numbers in the numerical code. This type of space filling curve is typically adopted by codes using the adaptive mesh refinement (Matsumoto 2007). The space-filling curve assigns the nodes whose MPI rank is close to the node number. In this section, the algorithm of the three-dimensional Peano-Hilbert space-filling curve is introduced.

First, we define the connection vectors for the first-order Peano-Hilbert curve as:

$$\mathbf{B_2} = (0, 1, 0), \tag{2.41}$$

$$\mathbf{B_3} = (1, 0, 0), \tag{2.42}$$

$$\mathbf{B_4} = (0, -1, 0), \tag{2.43}$$

$$\mathbf{B_5} = (0, 0, 1), \tag{2.44}$$

$$\mathbf{B_6} = (0, 1, 0), \tag{2.45}$$

$$\mathbf{B_7} = (-1, 0, 0), \tag{2.46}$$

$$\mathbf{B_8} = (0, -1, 0). \tag{2.47}$$

The first-order Peano-Hilbert curve has eight points that are connected with the connection vectors as follows:

$$^1\mathbf{P_1} = (0, 0, 0), \tag{2.48}$$

$$^1\mathbf{P_i} = {}^1\mathbf{P_{i-1}} + \mathbf{B_i} \quad \text{for } i = 2, 3, \ldots, 8. \tag{2.49}$$

The first-order Peano-Hilbert curve only consists of connection vectors. The first-order curve is shown in Fig. 2.2a.

Fig. 2.2 **a** First-order,
b second-order, and
c third-order Peano-Hilbert
space-filling curve. The
connection vectors are shown
in *red* (Color figure online)

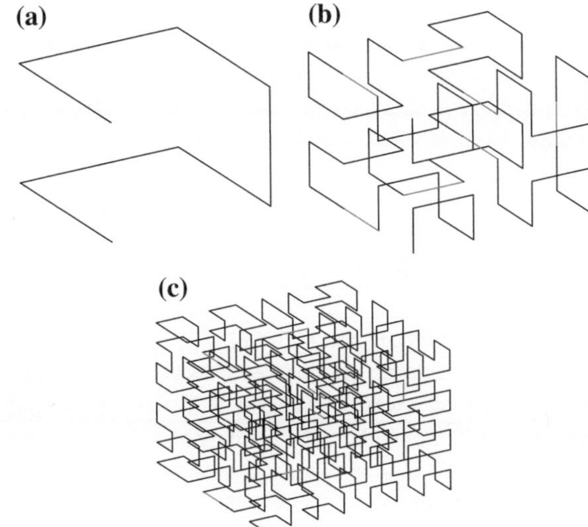

The second-order Peano-Hilbert curve is made by rotating and connecting the first-order curve. The rotation matrices are defined as

$$
\mathbf{A_1} = \begin{pmatrix} 1 & 0 & 0 \\ 0 & 0 & 1 \\ 0 & 1 & 0 \end{pmatrix}, \quad
\mathbf{A_2} = \begin{pmatrix} 0 & 0 & 1 \\ 0 & 1 & 0 \\ 1 & 0 & 0 \end{pmatrix},
$$

$$
\mathbf{A_3} = \begin{pmatrix} 1 & 0 & 0 \\ 0 & 1 & 0 \\ 0 & 0 & 1 \end{pmatrix}, \quad
\mathbf{A_4} = \begin{pmatrix} 0 & 0 & 1 \\ -1 & 0 & 0 \\ 0 & -1 & 0 \end{pmatrix},
$$

$$
\mathbf{A_5} = \begin{pmatrix} 0 & 0 & -1 \\ -1 & 0 & 0 \\ 0 & 1 & 0 \end{pmatrix}, \quad
\mathbf{A_6} = \begin{pmatrix} 1 & 0 & 0 \\ 0 & 1 & 0 \\ 0 & 0 & 1 \end{pmatrix},
$$

$$
\mathbf{A_7} = \begin{pmatrix} 0 & 0 & -1 \\ 0 & 1 & 0 \\ -1 & 0 & 0 \end{pmatrix}, \quad
\mathbf{A_8} = \begin{pmatrix} 1 & 0 & 0 \\ 0 & 0 & -1 \\ 0 & -1 & 0 \end{pmatrix}. \tag{2.50}
$$

Then, the eight rotated first-order curves are prepared as

$$
{}^1\mathbf{Q}_j^i = \mathbf{A_i}\,{}^1\mathbf{P_j}, \tag{2.51}
$$

Then the second-order curve is defined by connecting them as follows:

$$
{}^2\mathbf{P_i} = \begin{cases} {}^1\mathbf{Q}_i^1 & \text{for } i = 1, 2, \ldots, 8 \\ {}^2\mathbf{P_{i-1}} + \mathbf{B_n} & \text{for } m = 1 \; (i \neq 1), \\ {}^2\mathbf{P_{8(n-1)}} + {}^1\mathbf{Q}_m^n & \text{for others} \end{cases} \tag{2.52}
$$

where $m = i \bmod 8$ and $n = (i - m)/8 + 1$. The second order curve is shown in Fig. 2.2b. The connected vector is highlighted in red. Then, the same procedure is repeatedly applied and the higher-order curves are generated. The third-order curve is shown in Fig. 2.2c. We prepare the smallest Peano-Hilbert curve that covers all the nodes.

2.2.4 Yin-Yang Grid

To include the entire spherical shell, we adopt the Yin-Yang grid (Kageyama and Sato 2004). The Yin-Yang grid is a set of two congruent spherical geometries combined in a complementary way to cover the whole spherical shell. The boundary condition for each grid is calculated using the interpolation of the other grid. In all calculation, each grid covers $0.715R_\odot < r < r_{\max}, \pi/4 - \delta_\theta < \theta < 3\pi/4 + \delta_\theta$, and $-3\pi/4 - \delta_\phi < \phi < 3\pi/4 + \delta_\phi$, where δ_θ and δ_ϕ are the margins for interpolation. We use $\delta_\theta = 3\Delta\theta/2$ and $\delta_\phi = 3\Delta\phi/2$ for the interpolation with the third-order function, where $\Delta\theta$ and $\Delta\phi$ are the grid spacings in latitudinal and longitudinal directions. Although the boundary condition for the Yin grid is applied on the edge of the Yin grid, the boundary condition for the Yang grid is applied on the edge of the Yin grid to avoid the double solution in the overlapping area of the Yin-Yang grid (Fig. 2.3). The red thick lines show the location of the horizontal boundary for both Yin and Yang grids.

2.2.5 Big Data Management

The largest number of grid points in this thesis is $512(r) \times 1{,}024(\theta) \times 3{,}072(\phi) \times 2(\text{Yin Yang})$. For each output, the data is reorganized from the Yin-Yang grid to the ordinal spherical geometry with $512(r) \times 2{,}048(\theta) \times 4{,}096(\phi)$ grid points. With eight single-precision variables, it costs $512 \times 2{,}048 \times 4{,}096(\text{grid points}) \times 8(\text{variables}) \times 4(\text{single precision}) \sim 137\,\text{GB}$ per time step and $\sim 27\,\text{TB}$ for 200 steps. Thus, the treatment of such large number of data is no longer trivial. We adopt two strategies. First, the two-dimensional data in each layer with constant r is output to one file. Second, statistical values are calculated in the supercomputer and only the results that are likely smaller than the original are output. Figure 2.4 summarizes the first strategy. In the calculations, on account of the ordering using the Peano-Hilbert curve, the data distribution is rather complex which makes the analysis difficult. It is almost impossible to construct three-dimensional full data on a personal computer after the data transfer because of the computer memory requirements and reduction time. Thus, by using the first strategy, we can draw a two-dimensional map of the entire sphere with small tasks for download and reproduction. In the second strategy, the required statistical data are almost always the zonal average or at least the horizontal average, which can be generated from the layered data obtained by using the first

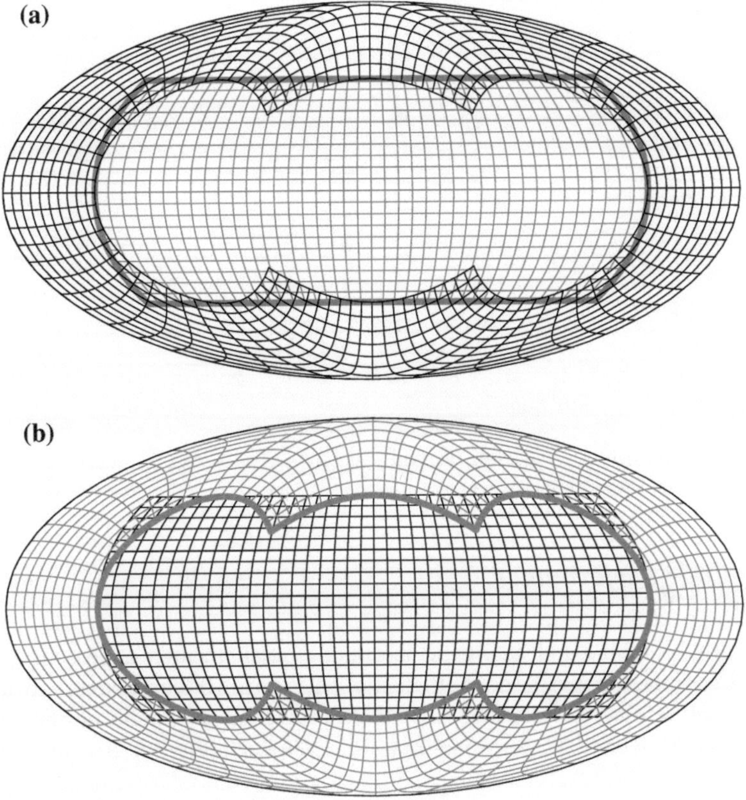

Fig. 2.3 *Red lines* and *Black lines* indicate the Yin and Yang grid, respectively. **a** and **b** shows the geometry on the Mollweide projection in the different view points. The *thick red lines* show the boundaries for both Yin and Yang grids (Color figure online)

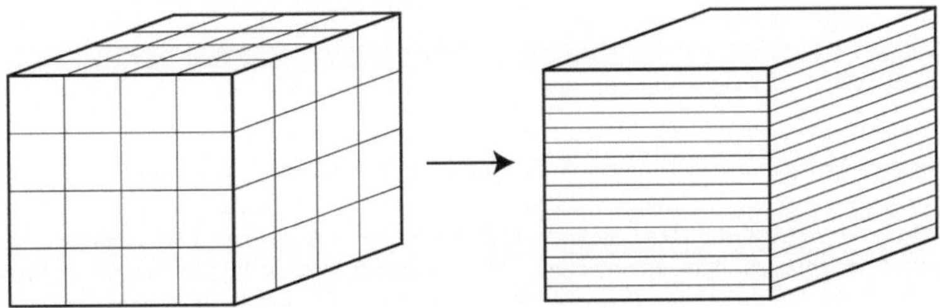

Fig. 2.4 Schematic for the first strategy. The complexly distributed data are reordered for analysis

strategy. We choose the required data in each two-dimensional layer, such as the RMS velocity, the energy flux and so on. This procedure significantly suppresses the required storage, analysis time, and the required memory a personal computer.

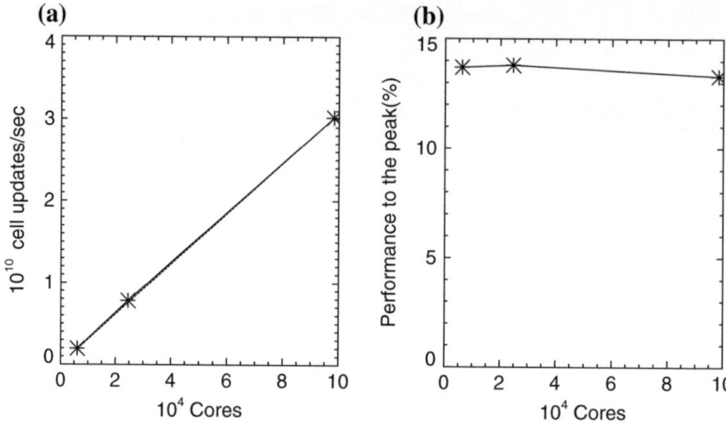

Fig. 2.5 **a** Performance for updating the grid point per second versus the number of cores and **b** peak floating point performance

2.2.6 Code Performance

Using a hybrid MPI and automatic intra-node parallelization approach, and the method explained above, the code efficiently scales up to 10^5 core counts with almost linear weak scaling and achieves 14 % performance at maximum on the RIKEN K-computer in Japan. The performance tests are shown in Fig. 2.5. Because the code includes almost no global communication among cores, this linear scaling is expected to hold further with larger cores. The code performs 3×10^5 grid update/core/s, which allows to investigate the interaction of small-scale and large-scale convection in the spherical shell.

References

J. Christensen-Dalsgaard, W. Dappen, S.V. Ajukov, E.R. Anderson, H.M. Antia, S. Basu, V.A. Baturin, G. Berthomieu, B. Chaboyer, S.M. Chitre, A.N. Cox, P. Demarque, J. Donatowicz, W.A. Dziembowski, M. Gabriel, D.O. Gough, D.B. Guenther, J.A. Guzik, J.W. Harvey, F. Hill, G. Houdek, C.A. Iglesias, A.G. Kosovichev, J.W. Leibacher, P. Morel, C.R. Proffitt, J. Provost, J. Reiter, E.J. Rhodes Jr., F.J. Rogers, I.W. Roxburgh, M.J. Thompson, R.K. Ulrich, The current state of solar modeling. Science **272**, 1286–1292 (1996). doi:10.1126/science.272.5266.1286
H. Hotta, M. Rempel, T. Yokoyama, Y. Iida, Y. Fan, Numerical calculation of convection with reduced speed of sound technique. A&A **539**, 30 (2012). doi:10.1051/0004-6361/201118268
A. Kageyama, T. Sato, "Yin-Yang grid": An overset grid in spherical geometry. Geochem. Geophys. Geosyst. **5**, 9005 (2004). doi:10.1029/2004GC000734
T. Matsumoto, Self-gravitational magnetohydrodynamics with adaptive mesh refinement for protostellar collapse. PASJ **59**, 905 (2007)

M.S. Miesch, J.R. Elliott, J. Toomre, T.L. Clune, G.A. Glatzmaier, P.A. Gilman, Three-dimensional spherical simulations of solar convection. I. Differential rotation and pattern evolution achieved with laminar and turbulent states. ApJ **532**, 593–615 (2000). doi:10.1086/308555

D. Mihalas, B.W. Mihalas, *Foundations of Radiation Hydrodynamics* (Oxford University Press, New York, 1984)

M. Rempel, M. Schüssler, M. Knölker, Radiative magnetohydrodynamic simulation of sunspot structure. ApJ **691**, 640–649 (2009). doi:10.1088/0004-637X/691/1/640

F.J. Rogers, F.J. Swenson, C.A. Iglesias, OPAL equation-of-state tables for astrophysical applications. ApJ **456**, 902 (1996). doi:10.1086/176705

R.F. Stein, Å. Nordlund, D. Georgoviani, D. Benson, W. Schaffenberger, Supergranulation-scale convection simulations, in *Solar-Stellar Dynamos as Revealed by Helio- and Asteroseismology: GONG 2008/SOHO 21*. Astronomical Society of the Pacific Conference Series, vol. 416, ed. by M. Dikpati, T. Arentoft, I. González Hernández, C. Lindsey, F. Hill (2009), p. 421

A. Vögler, S. Shelyag, M. Schüssler, F. Cattaneo, T. Emonet, T. Linde, Simulations of magneto-convection in the solar photosphere. Equations, methods, and results of the MURaM code. A&A **429**, 335–351 (2005). doi:10.1051/0004-6361:20041507

Chapter 3
Structure of Convection and Magnetic Field Without Rotation

Abstract Using the developed numerical code, we perform non-rotating high-resolution calculations of solar global convection, which resolve convective scales of less than $10\,\mathrm{Mm}$. The main conclusions of this study are the following. (1) The small-scale downflows generated in the near surface layer penetrate down to deeper layers and excite small-scale turbulence in the region of $>0.9R_\odot$, where $R_\odot$ is the solar radius. (2) In the deeper convection zone ($<0.9R_\odot$), the convection is not affected by the location of the upper boundary. (3) Using an LES (Large Eddy Simulation) approach we achieved small-scale dynamo action and maintained a field of $0.15 - 0.25B_\mathrm{eq}$ throughout the convection zone, where B_eq is the equipartition magnetic field to the kinetic energy. (4) The overall dynamo efficiency significantly varies in the convection zone as a consequence of the downward directed Poynting flux and the depth variation in the intrinsic convective scales. For a fixed numerical resolution the dynamo relevant scales are better resolved in the deeper convection zone and are therefore less affected by numerical diffusivity, i.e. the effective Reynolds numbers are larger.

Keywords Small-scale dynamo · Large eddy simulation · Reduced speed of sound technique

3.1 Introduction

The purpose of the study in this chapter is achieving multi-scale convection including scales from $10\,\mathrm{Mm}$ (smaller than supergranulation) up to $\sim 200\,\mathrm{Mm}$ (global scale). We achieve this by approaching the solar surface up to $0.99R_\odot$ using unprecedented resolution and study in particular the influence of the location of the upper boundary on the convective structure in the deeper parts of the convection zone. In addition we use our setup to study the transport and the generation of magnetic field by turbulent convection in the absence of rotation, i.e., we study the operation of a small-scale dynamo in the highly stratified convection zone.

The chapter is organized as follows: We introduce our numerical setting in Sect. 3.2 and show the result in Sect. 3.3: The properties based on the obtained spatial

© Springer Japan 2015

H. Hotta, *Thermal Convection, Magnetic Field, and Differential Rotation in Solar-type Stars*, Springer Theses, DOI 10.1007/978-4-431-55399-1_3

distributions are shown in Sect. 3.3.1. The discussion about the energy balance using the integrated flux is given in Sect. 3.3.2. In Sect. 3.3.3, the analysis for the properties of the convection in cases without the magnetic fields using the spherical harmonic expansion is shown. Then the analysis for the cases with magnetic fields using spherical harmonic expansion and the probability density function are shown in Sect. 3.3.4. The investigation of the transportation and the generation of the magnetic field in the convection zone is done in Sect. 3.3.5. In Sect. 3.4, we summarize this chapter.

3.2 Model

In this chapter the rotation is not included, since we focus on the connection of thermal convection between the small and large scales and its local dynamo action in the global scale. The rotation with the thermal convection generates differential rotation and this causes the global dynamo action, such as the Ω-effect. It is difficult to distinguish the local dynamo action caused by turbulence and the global dynamo caused by the global mean flow. The calculation with the rotation is shown in Chap. 4. We use the stress-free and the inpenetrative boundary condition for the fluid velocity, v_r, v_θ, and v_ϕ. The free boundary condition is adopted for the density and the entropy. The magnetic field is vertical at the top boundary and the perfect conductor boundary condition is used at the bottom boundary.

ρ_1, B_r, B_θ, B_ϕ and s_1 are zero initially. The fluid velocities v_r, v_θ, and v_ϕ has small random values. After the convection reaches statistically steady state, the uniform magnetic field ($B_\phi = 100\,\mathrm{G}$) is added. Although the net flux exists initially using this condition, this disappears through upper boundary around 75 days at all. Thus, we do not have to consider the influence of the net flux for our local dynamo study.

We carry out three calculations named H0, H1 and H2, with the different setting which is firstly hydrodynamic (see the Table 3.1). In the case H0, the top boundary locates at $r = 0.99R_\odot$ and the density contrast $\rho_0(r_{\min})/\rho_0(r_{\max})$ exceeds 600. To our knowledge, this is the largest value in the numerical calculation achieved so far in numerical simulation of solar global convection. In the cases H1 and H2, the top boundary is at $r = 0.96R_\odot$ and the density contrast is around 40. In the case H1, the thickness of the cooling layer is the same as the case H0, i.e. $d_c = 3,740\,\mathrm{km}$ which is the two pressure scale heights at $r = 0.99R_\odot$, while the case H2 adopt two scale

Table 3.1 Important parameters in our studies

Case	H0	H1	H2
N_r	512	456	456
$r_{\max}/R_\odot$	0.99	0.96	0.96
$\dfrac{\rho_0(r_{\min})}{\rho_0(r_{\max})}$	613	36	36
$H_{p0}(r_{\max})$ (km)	1,870	9,390	9,390
d_c (km)	3,740	3,740	18,780

heights at its top boundary ($r_{\mathrm{max}} = 0.96R_\odot$) for the thickness ($d_c$: see Chap. 3). After the uniform magnetic field is added, the calculations are newly named M0, M1 and M2, which use the results of H0, H1, and H2, respectively. We note that the thickness of the cooling layer (d_c) has almost the same role as the value of the turbulent thermal diffusivity on the entropy that is adopted in ASH simulation (Miesch et al. 2000, 2008). After the uniform magnetic field is added the calculations are newly named M0, M1 and M2, which use the results of H0, H1, and H2, respectively.

The horizontal grid spacing is 1,100 km at the top boundary and radially 375 km. Using the Yin-Yang geometry, the number of grid points is $1,024(N_\theta) \times 3,072$ (N_ϕ) $\times$ 2. The last factor 2 indicates a pair of the Yin and Yang. The number of grid points in the radial direction is shown in the Table 3.1. Since this resolution in the Yin-Yang geometry has almost the same quality as that of $512(r) \times 2,048(\theta) \times 4,096(\phi)$ in ordinary spherical geometry in the case H0 and M0, we succeed in doubling the resolution in each direction from the previous study (Miesch et al. 2008).

3.3 Results

3.3.1 Structure of Convection and Magnetic Field

Figure 3.1 shows the RMS velocities in the case H0, H1, and H2. The maximum RMS velocity is 5×10^4 cm s^{-1} at the top boundary in the case H0. Since the Mach

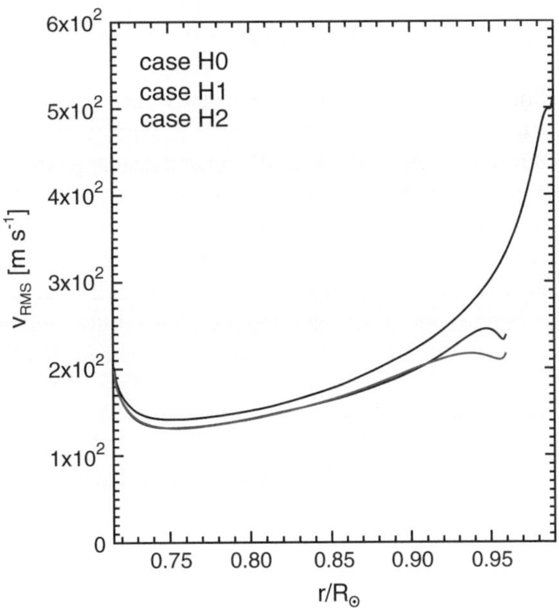

Fig. 3.1 RMS (root mean square) velocities as a function of the depth are shown. The *black*, *blue*, and *red lines* show the results in the case H0, H1, and H2, respectively (Color figure online)

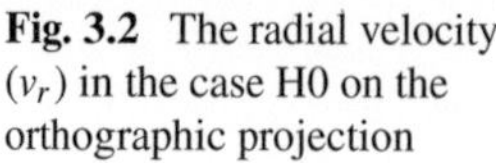

Fig. 3.2 The radial velocity
(v_r) in the case H0 on the
orthographic projection

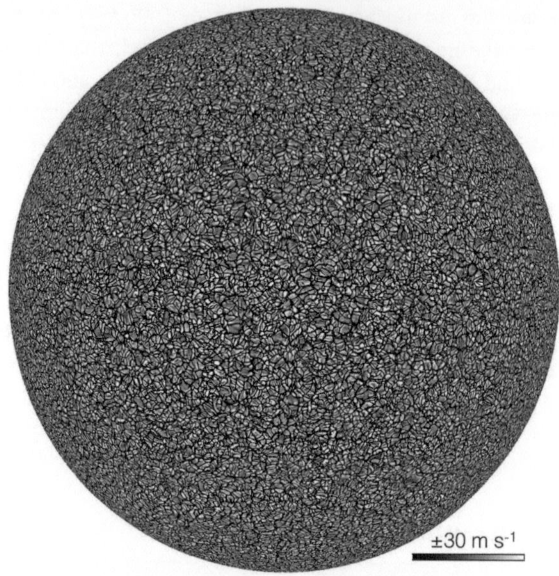

number determined with the reduced speed of sound $\hat{c}_s$ is always under 0.3 all over
the convection zone, the requirement for the validity of the RSST in this study is
well satisfied (Hotta et al. 2012b).

Figure 3.2 shows the radial velocity (v_r) around the top boundary in the case H0
($r_{max} = 0.99R_\odot$, which is 7 Mm below the photosphere) in the orthographic projection. Note that since the shown location is close to the impenetrable top boundary,
the value of the radial velocity is rather small. Since near the upper boundary the
pressure scale height is less than 2 Mm, the convection pattern shows small cells of
about ($\sim$7 Mm). The typical cell size is slightly smaller than supergranulation that is
observed on the photosphere. This is the first work that well resolves the 10 Mm-scale
convection pattern in a calculation of the solar global convection zone. Figure 3.3a–c
shows the radial velocity at $r = 0.99R_\odot$, $r = 0.95R_\odot$, and $r = 0.85R_\odot$ in the case
H0 by using the orthographic projection. In deeper layer the pressure scale height
increases and the convection pattern becomes larger. The detailed analysis using the
spherical harmonic expansion of the convective structure is shown in Sect. 3.3.3.

Figure 3.4a, d, and g show the zoomed-up contour of the radial velocity in the
case M0, i.e., after the inclusion of the magnetic field in H0. The region is indicated
by the white rectangle in Fig. 3.3a–c. Figure 3.4c, f, and i show the vorticity ($\omega_r =
(\nabla \times \mathbf{v})_r$) at $r = 0.99R_\odot$, $0.95R_\odot$, and $0.85R_\odot$, respectively. As already seen in
the case H0 (Fig. 3.3), it is clear that the scale of the thermal convection pattern
significantly depends on the depth. In addition, the large-scale downflow is associated
with small-scale and strong vorticity in the deeper layer (especially in $r = 0.85R_\odot$).
Figure 3.5a shows $\rho_0[s_1(r, \theta, \phi = 0) - \langle s_1 \rangle]$ in the meridional plane, where $\langle\rangle$ shows
the horizontal average in this chapter. The low and high entropy materials correspond
to the downflow and upflow, respectively. In the near surface region, the convection

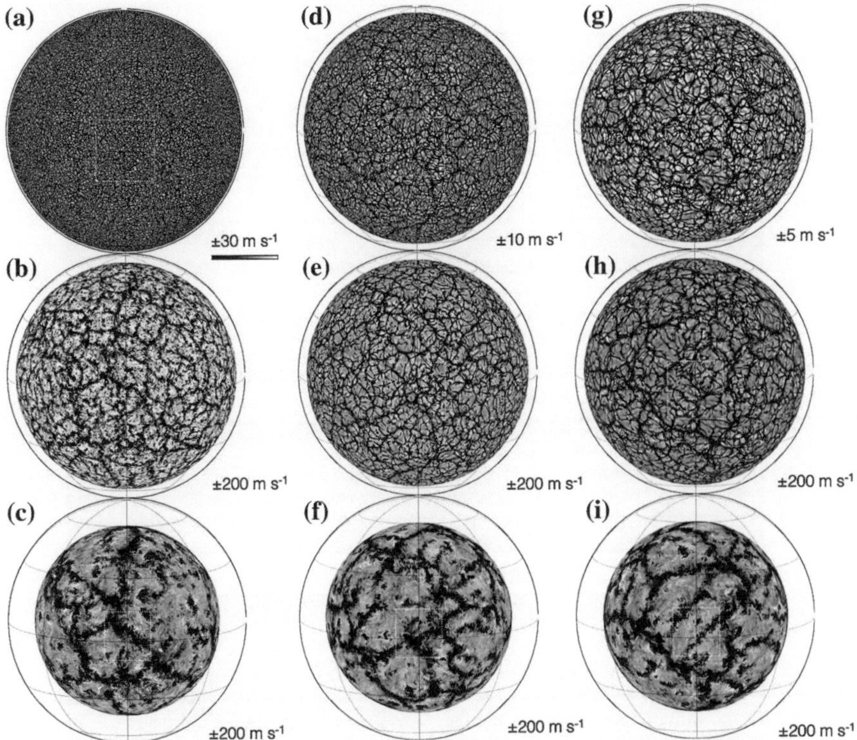

Fig. 3.3 The radial velocities (v_r) are shown at $r = r_{\max}$ (**a, d, g**), $r = 0.95R_\odot$ (**b, e, h**) and $r = 0.85R_\odot$ (**c, f, i**). The results in the cases H0, H1, and H2 are shown in (**a–c**), (**d–f**), and (**g–i**), respectively. The *black circle* around each panel show the location at $r = R_\odot$

structure shows a combination of broad upflows and narrow downflows $\sim$7 Mm forming at the top boundary. These small-scale downflows merge in the middle of the convection zone and build large-scale downflow. Although the overall structure of such convection is large, there is a superimposed turbulent pattern especially in the downflow region which is shown in Fig. 3.4.

Figure 3.3d–f show the results of the case H1 with different location of the top boundary ($r_{\max} = 0.96R_\odot$) at $r = r_{\max}$, $r = 0.95R_\odot$, and $r = 0.85R_\odot$, respectively. Since the location and the pressure scale height of shown images is different between the case H0 in Figs. 3.3a ($r = r_{\max} = 0.99R_\odot$) and H1 in d ($r = r_{\max} = 0.96R_\odot$), it is natural that the convective structures are much different, i.e., those in H0 has smaller scale convection than in H1. It is more important that the structures at the same depth, $0.95R_\odot$, are significantly different with each other between these cases (H0 in Fig. 3.3b and H1 in Fig. 3.3e). The small-scale downflow plumes penetrate near surface layer and influence its structure (see also Fig. 3.5a). When the downflow goes deeper, the influence becomes smaller. This is seen in comparison of Fig. 3.3c ($r_{\max} = 0.99R_\odot$ and $r = 0.85R_\odot$) and f ($r_{\max} = 0.96R_\odot$ and $r = 0.85R_\odot$), where the difference of convection structure seems insignificant. Figure 3.3g–i show the

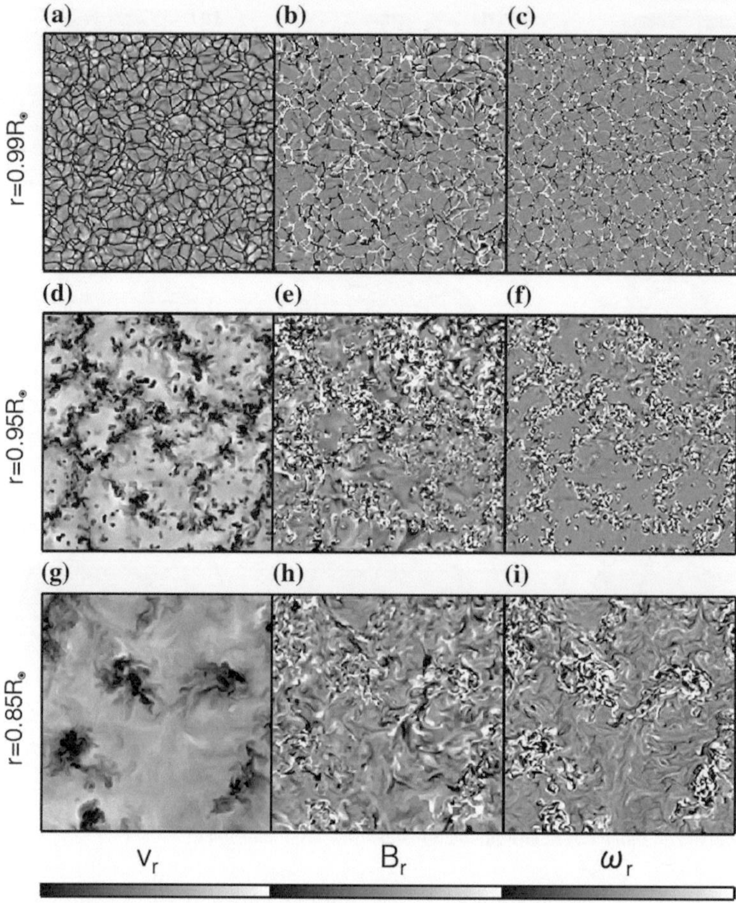

Fig. 3.4 The zoomed-in contour of the radial velocity (v_r: *left panels*), the radial magnetic field (B_r: *middle panels*) and the radial vorticities (ω_r: *right panels*) in different depth. The field of view is 30° both in the latitude and the longitude that corresponds to the size of 370 Mm at the *top* boundary. The result is from the case H0, **a** $\pm60\,(\mathrm{m\,s^{-1}})$, **b** $\pm50\,(\mathrm{G})$, **c** $\pm1 \times 10^{-4}\,(\mathrm{s^{-1}})$, **d** $\pm300\,(\mathrm{m\,s^{-1}})$, **e** $\pm500\,(\mathrm{G})$, **f** $\pm1 \times 10^{-4}\,(\mathrm{s^{-1}})$, **g** $\pm300\,(\mathrm{m\,s^{-1}})$, **h** $\pm2000\,(\mathrm{G})$, **i** $\pm4 \times 10^{-5}\,(\mathrm{s^{-1}})$

results in the case H2 in which the location of the top boundary is the same as H1 but the thickness of the cooling layer is larger. The convective structure around the top boundary shows the largest scale (Fig. 3.3g), while again the difference becomes smaller in the deeper layer (Fig. 3.3c, f, and i). This is also shown in Fig. 3.6 by comparing the occupied area by the upflow and the downflow, i.e., positive and negative radial velocities (v_r). Up to the middle of the convection zone ($<0.9R_\odot$), all the cases H0, H1 and H2 show similar behavior, i.e., fractional area by the upflow is larger ($\sim65\%$) but decreases below $\sim0.85R_\odot$ to equal to that of the downflow. Interestingly, this behavior is quantitatively the same in spite of significant difference between H0 and H1 in the density contrast (see Table 3.1).

Fig. 3.5 **a** $\rho_0[s_1 - \langle s_1 \rangle]$, and
b $B/\sqrt{4\pi\rho_0}$ on the
meridional plane at $\phi = 0$,
where $\langle s_1 \rangle$ is the horizontal
average of the entropy. The
result is from the case H0

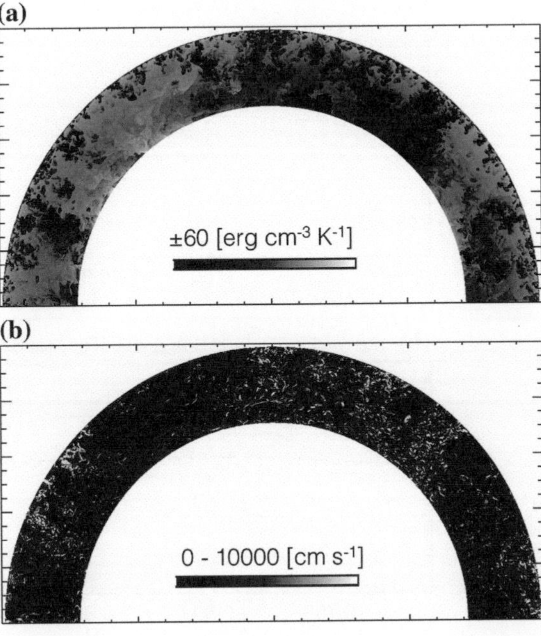

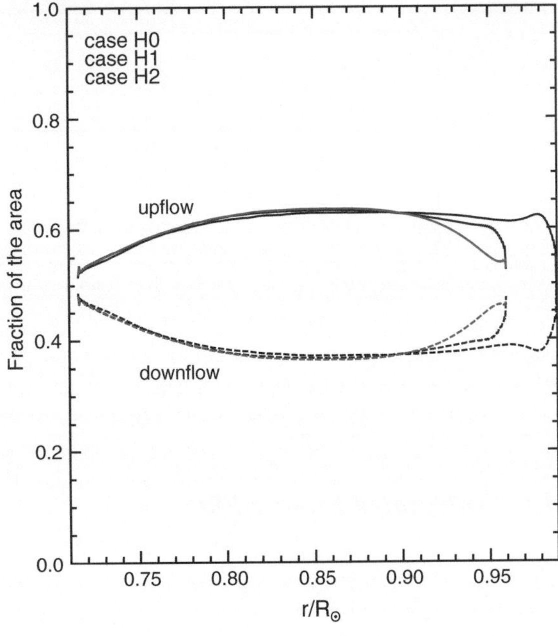

Fig. 3.6 The occupied
fraction of area of upflow
(*solid line*) and downflow
(*dashed line*). The *black,
blue* and *red lines* show the
results in the case H0, H1,
and H2, respectively (Color
figure online)

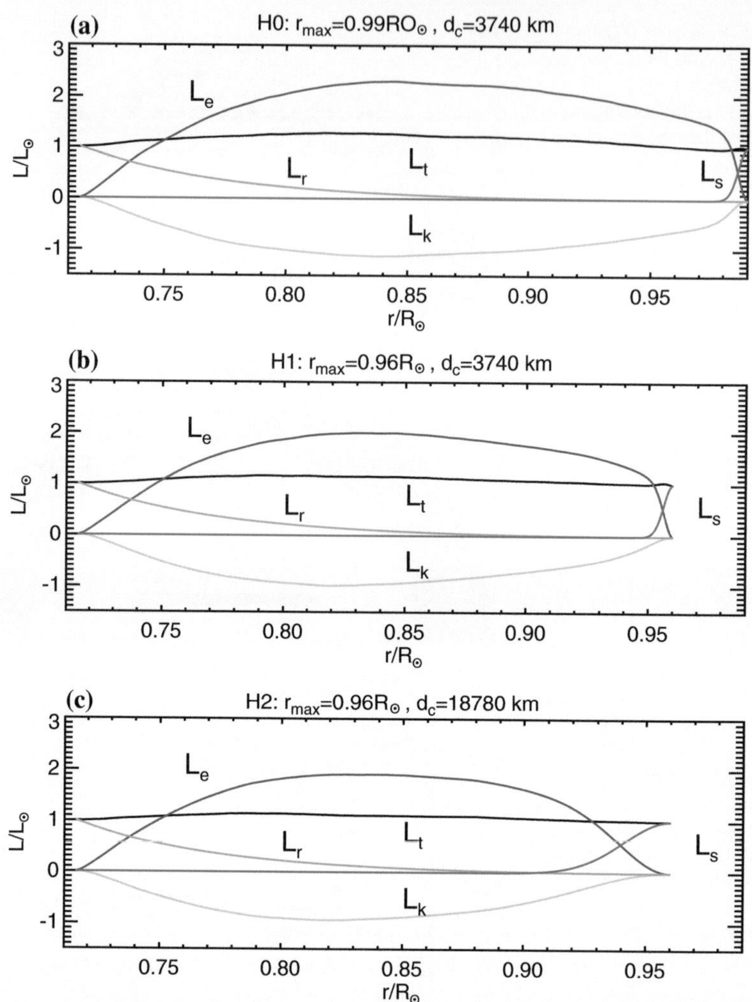

Fig. 3.7 The integrated fluxes are shown. The panels **a, b,** and **c** show the result in the cases H0, H1, and H2, respectively. The *black, blue, green, red,* and *light blue lines* show the total, enthalpy, radiative, surface cooling, and kinetic fluxes, respectively (Color figure online)

3.3.2 Integrated Energy Flux

Figure 3.7 shows the integrated fluxes. The integrated enthalpy flux (L_e), the integrated kinetic flux (L_k), the radiative luminosity (L_r) and the luminosity form of the surface cooling (L_s) are defined as

$$L_e = \int_S \left[\rho_0 e_1 + p_1 - \frac{p_0 \rho_1}{\rho_0} \right] v_r \, dS, \tag{3.1}$$

$$L_k = \int_S \frac{1}{2} \rho_0 v^2 v_r \, dS, \tag{3.2}$$

$$L_r = \int_S F_r(r) \, dS, \tag{3.3}$$

$$L_s = \int_S F_s(r) \, dS. \tag{3.4}$$

The radiative flux (F_r) and surface cooling flux (F_s) are defined in this chapter. Figure 3.7a shows the integrated fluxes in the case H0 ($r_{max} = 0.99 R_\odot$ and $d_c = 3{,}740\,$km). The total integrated flux $L_t = L_e + L_k + L_r + L_s$ is almost constant along the depth. This indicates that the convection zone in our calculation is in the energy equilibrium. Note that since we do not use a conservative form for the total energy nor estimate the energy flux contributions caused by the artificial diffusivity, the total flux is not completely constant. The enthalpy flux transports twice the solar luminosity upward at maximum and the kinetic flux transports almost the same amount of the energy as the solar luminosity downward at maximum. Although the kinetic energy flux is frequently ignored in the one-dimensional mixing length model (e.g. Stix 2004), our result shows the importance of the kinetic flux. This has been already suggested by Miesch et al. (2008). Figure 3.7b, and c show the results in the case H1 and H2, respectively. The integrated fluxes show almost the similar behavior as those in the case H0, but the maximum absolute values of the enthalpy and kinetic flux are smaller. Since these absolute values gradually decrease from H0 to H2, we conclude that both the thickness of the cooling layer and the location of the upper boundary contribute to this issue. It is possible that, when the upper boundary becomes closer to the real solar surface and the cooling layer becomes thinner, the absolute values of the enthalpy flux and kinetic flux become even larger than our case H0.

Figure 3.8 shows the integrated enthalpy flux and kinetic flux transported by upflow (L_{eu}, L_{ku}) and downflow (L_{ed}, L_{kd}), respectively. Note that the enthalpy flux by upflow and downflow is estimated with the perturbation from the reference state. Regarding the enthalpy flux, both upflow and downflow transport energy upward. The downflow transports most of energy ($>70\,\%$). We note that the enthalpy flux of the upflow shows the negative value near the bottom boundary, since the cool fluid bounces at the boundary and moves upward. Regarding the kinetic energy flux, upflows (downflows) transport energy upward (downward). The larger kinetic energy flux of the downflow makes the kinetic energy flux negative. These results show that downflows play a key role in the transports of energy in the solar convection zone. While we find a significant difference in the overall amplitude among cases H0–H2, the ratio of contributions from up- and downflows does not change much despite the significant difference in the density contrast.

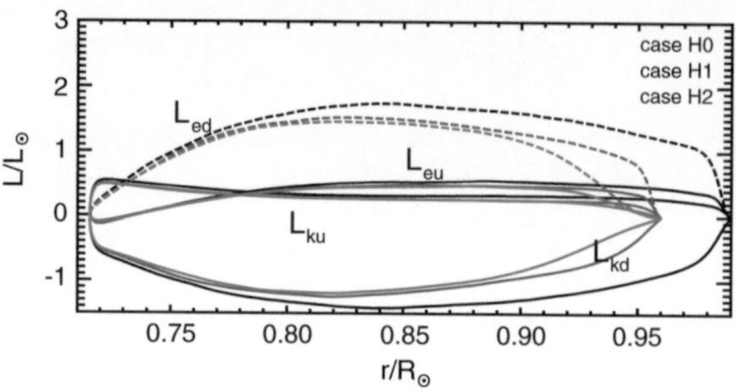

Fig. 3.8 The enthalpy and kinetic flux transported by the upflow (L_{eu} and L_{ku}) and the downflow (L_{ed} and L_{eu}) are shown. The *black, blue* and *red lines* show the result in the cases H0, H1, and H2, respectively. The *solid* and *dashed lines* indicate the enthalpy flux in upflow and downflow respectively and the *dash-dot* and *dash-dot-dot-dot lines* indicate the kinetic energy flux in upflow and downflow respectively (Color figure online)

3.3.3 Analysis Using Spherical Harmonics for Hydrodynamic Cases

In this section, the results of the analysis using the spherical harmonics expansion are shown. We focus on the question: How do the location of the upper boundary and the thickness of the surface cooling layer influence the convective structure throughout the convection zone?

A real function $f(\theta, \phi)$ can be expressed in spherical harmonics as

$$f(\theta, \phi) = \sum_{l=0}^{l_{max}} \sum_{m=-l}^{l} f_{lm} Y_{lm}(\theta, \phi), \tag{3.5}$$

where $Y_{lm}(\theta, \phi)$ is the spherical harmonics for degree l and order m. In the analyses the absolute total value along m without $m = 0$

$$\bar{f}_l = \sqrt{\sum_{m=-l}^{l} |f_{lm}|^2}, \tag{3.6}$$

is shown. The value is normalized in order to satisfy the relation:

$$\frac{\int_\Omega (f(\theta, \phi) - \langle f(\theta, \phi) \rangle)^2 \sin\theta d\theta d\phi}{4\pi} = \sum_{l=1}^{l_{max}} \bar{f}_l^2. \tag{3.7}$$

Our spherical harmonic analyses are performed using the freely available software archive SHTOOLS (shtools.ipgp.fr).

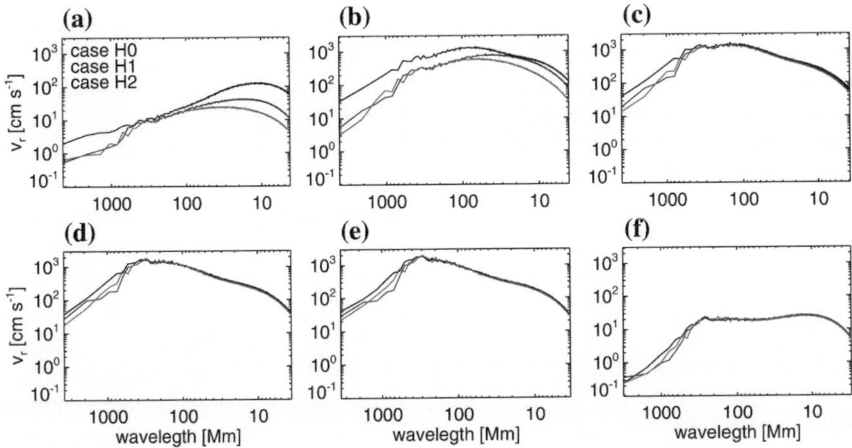

Fig. 3.9 Spectra of the radial velocity at **a** $r = r_{max}$, **b** $r = 0.95R_\odot$, **c** $r = 0.90R_\odot$, **d** $r = 0.85R_\odot$, **e** $r = 0.80R_\odot$, **f** $r = 0.715R_\odot$. The *black, blue,* and *red lines* specify the results in the case H0, H1, and H2, respectively (Color figure online)

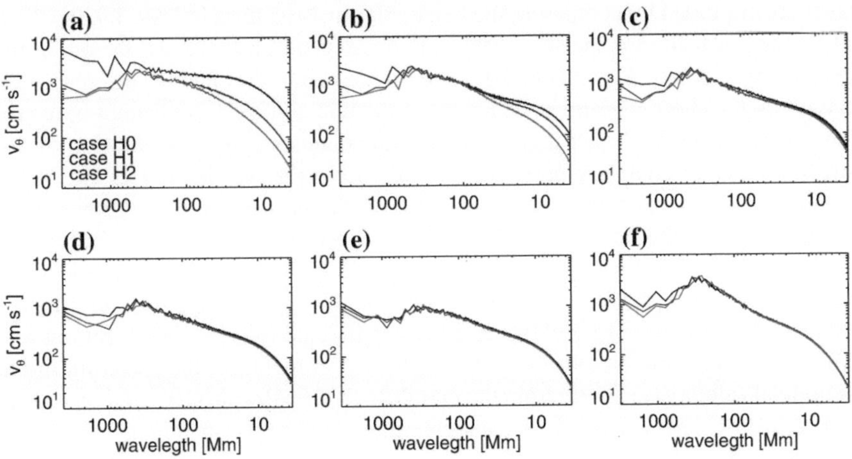

Fig. 3.10 Spectra of the latitudinal velocity at **a** $r = r_{max}$, **b** $r = 0.95R_\odot$, **c** $r = 0.90R_\odot$, **d** $r = 0.85R_\odot$, **e** $r = 0.80R_\odot$, **f** $r = 0.715R_\odot$. The *black, blue,* and *red lines* specify the results in the case H0, H1, and H2, respectively (Color figure online)

Figures 3.9 and 3.10 show the spectra of the radial velocity (v_r) and the latitudinal velocity (v_θ), respectively as a function of the horizontal wavelength ($L_h = 2\pi r / l$), where l is the spherical harmonic degree, i.e., the horizontal wavenumber. The black, blue, and red lines show the results in the case H0 ($r_{max} = 0.99R_\odot$ and $d_c = 3,740$ km), H1 ($r_{max} = 0.96R_\odot$ and $d_c = 3,740$ km), and H2 ($r_{max} = 0.96R_\odot$ and $d_c = 18,780$ km), respectively. The black line in Fig. 3.9a shows a peak around $L_h \sim 7$–8 Mm (see also Fig. 3.2). This peak moves to the larger scale L_h, with increasing depth. At $r = 0.80R_\odot$, the peak is around $L_h \sim 300\,\mathrm{Mm}$ (Fig. 3.9f, black line). This reflects the variation of the pressure scale height, in the solar model, $H_p = 1.9\,\mathrm{Mm}$ at $r = 0.99R_\odot$ and $H_p = 44\,\mathrm{Mm}$ at $r = 0.8R_\odot$. Again the peak is in the smaller scale ($\sim7\,\mathrm{Mm}$) at the bottom boundary $r = r_{\min} = 0.715R_\odot$. This is caused by the collision between the downflow and the impenetrable bottom boundary. In this process a thin boundary layer whose thickness is determined by the numerical resolution is formed. The scale of turbulence near the boundary (Fig. 3.9f) is a consequence of this boundary layer. Comparison of black lines in Figs. 3.9a and 3.10a indicates that the peak of the latitudinal velocity is at a much larger scale ($\sim400\,\mathrm{Mm}$) than the radial velocity even close to the upper boundary. We note that almost all the spectra have this feature around $L_h \sim 400\,\mathrm{Mm}$, i.e., the peak or the bended feature in which power-law index varies. The scale $400\,\mathrm{Mm}$ is the length twice the thickness of the convection zone, i.e., the radial extent of our computational domain. This result is consistent with our previous study (Hotta et al. 2012a) which argues that the typical scale of the convection is determined by the pressure (or density) scale height or the height of the computational domain. Figure 3.10a shows that the horizontal velocity is more likely to be influenced by the large-scale structure.

The influences from the location of the boundary and the thickness of the cooling layer are discussed by comparing the black, blue and red lines in Figs. 3.9 and 3.10. The common feature is that the spectra do not depend on these two factors in the deep layer ($r \leq 0.85R_\odot$: panels d–f). This suggests that the small-scale convection caused by the short pressure scale height or the thin cooling layer cannot influence the convection in the deeper region. On the top boundary (Figs. 3.9a and 3.10a), the small-scale convection is suppressed gradually from the case H0 to H2. This is the confirmation of our understanding from the appearance of the convection pattern in Sect. 3.3.1. In the near surface layer ($r = 0.95R_\odot$), the difference still remains unchanged. While the upper location of the boundary increases the amplitude of the fluid velocity in all the scale (the black line), the thin cooling layer excites only the small-scale convection ($<50\,\mathrm{Mm}$) and the larger scale remains. We point out that when the surface layer ($0.96R_\odot < r < 0.99R_\odot$) is included, the spectrum of the latitudinal velocity (v_θ) is flat from the middle to the small scale ($10 < L_h < 40\,\mathrm{Mm}$ in Fig. 3.10b). This is caused by the penetration of the corresponding-scale plume from the near surface layer.

3.3.4 Analysis Using Spherical Harmonics and Probability Density Function for Magnetohydrodynamic Cases

In this subsection, we analyze the results of the magnetohydrodynamic calculation in the cases M0, M1, and M2 using the spherical harmonics and the probability density function. We focus on the influence of the location of the boundary layer and the thickness of the cooling layer on the structure of the magnetic field.

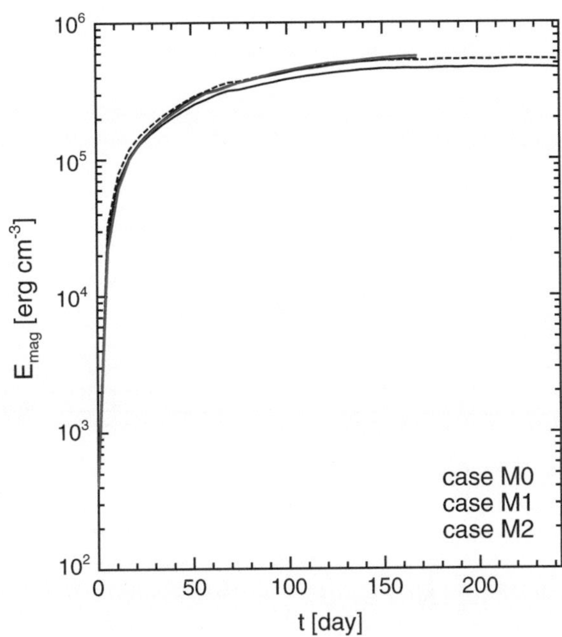

Fig. 3.11 The time evolution of the average magnetic energy is shown. The *black, blue* and *red lines* show the result in the case M0, M1 and M2, respectively. The *dashed line* shows the magnetic energy in the case M0 averaged over $r_{\min} < r < 0.96R_\odot$. $t = 0$ is time at which the magnetic field is inputted (Color figure online)

Figure 3.11 shows the time evolution of the magnetic energy ($B^2/(8\pi)$) averaged over the simulation domain at each time step. The initial linear growth stops around 10 days after the input of the seed field in every case. Even after that the magnetic energy continues to increase gradually with a rather small growth rate until it saturates after around 150 days in all cases. The comparison between cases is done using the data from $t = 115$ day to $t = 162$ day in which the generation of the magnetic field is almost saturated. Basically the differences between the cases are insignificant, although the case M0 saturates with a slightly smaller average magnetic energy. Since the equipartition magnetic field strength in the near surface region is smaller due to the small density (ρ_0), the increase of the volume in the case M0 causes the slight decrease in the average magnetic energy. This can be supported from the fact that average value over $r_{\min} < r < 0.96R_\odot$ in M0 is larger as shown by the dashed line in Fig. 3.11. This indicates higher growth rate of the average magnetic energy there than those in cases M1 and M2. The time scale of the convection in the near surface layer is short. The generated magnetic field is transported downward (see also the discussion about the pumping in Sect. 3.3.5).

Figure 3.12 shows the spectra of the magnetic energy $B^2/(8\pi)$ at selected depth. Similar to the results introduced in the previous sections, the difference can be seen in the near surface layer, and this difference becomes insignificant as we go to the deeper layers.

Figure 3.13 shows the probability density functions (PDFs) for the three components of velocity (v_r, v_θ, and v_ϕ) and the three components of magnetic field (B_r, B_θ, and B_ϕ), the radial vorticity ω_r, the horizontal divergence ζ and the temperature

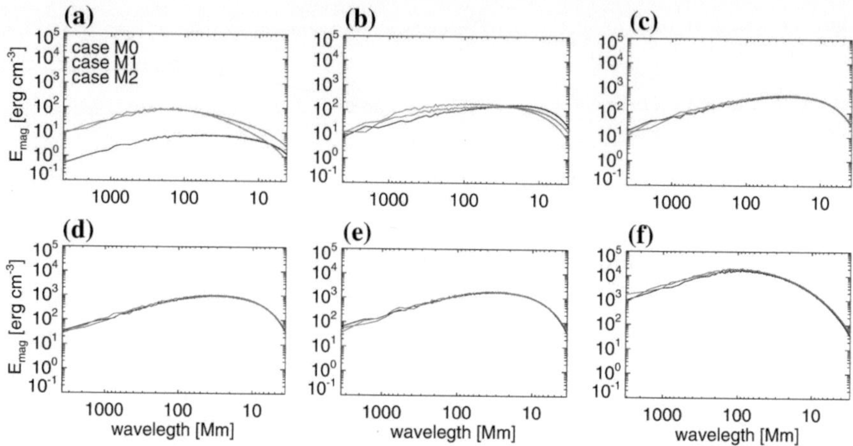

Fig. 3.12 Spectra of the magnetic energy ($B^2/8\pi$) at **a** $r = r_{\max}$, **b** $r = 0.95R_\odot$, **c** $r = 0.90R_\odot$, **d** $r = 0.85R_\odot$, **e** $r = 0.80R_\odot$, **f** $r = 0.715R_\odot$. The *black, blue,* and *red lines* specify the results in the case M0, M1, and M2, respectively (Color figure online)

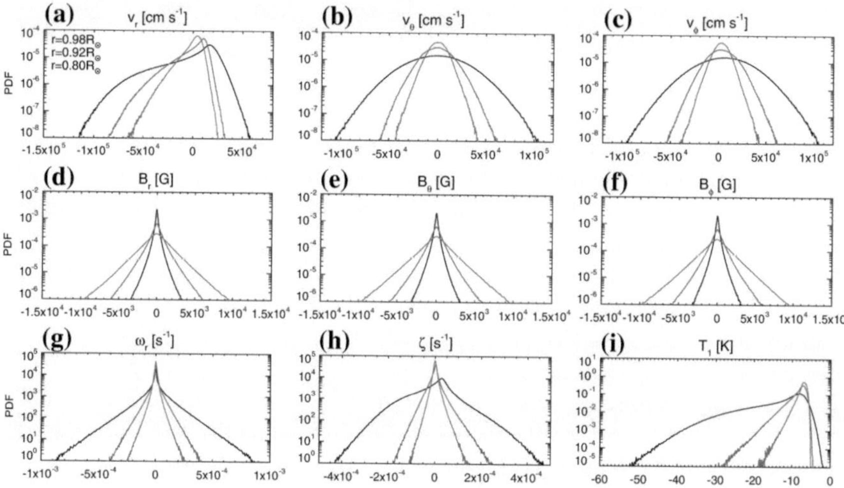

Fig. 3.13 The probability density function of **a** the radial velocity, **b** the latitudinal velocity, **c** the longitudinal velocity, **d** the radial magnetic field, **e** the latitudinal magnetic field, **f** the longitudinal magnetic field, **g** the radial vorticity, **h** the horizontal divergence, and **i** the temperature perturbation are shown using the result in the case M0 at $t = 115$ day (Color figure online)

perturbation T_1 in the case M0 at $t = 115$ day. Although the rotation is not included in this study, some features of the PDFs are similar to findings from previous studies including rotation (Brun et al. 2004; Miesch et al. 2008). In this study the PDF is the normalized histogram on a horizontal surface, corrected for the grid convergence at the poles. Figure 3.13a shows the significant asymmetry in the radial velocity. This

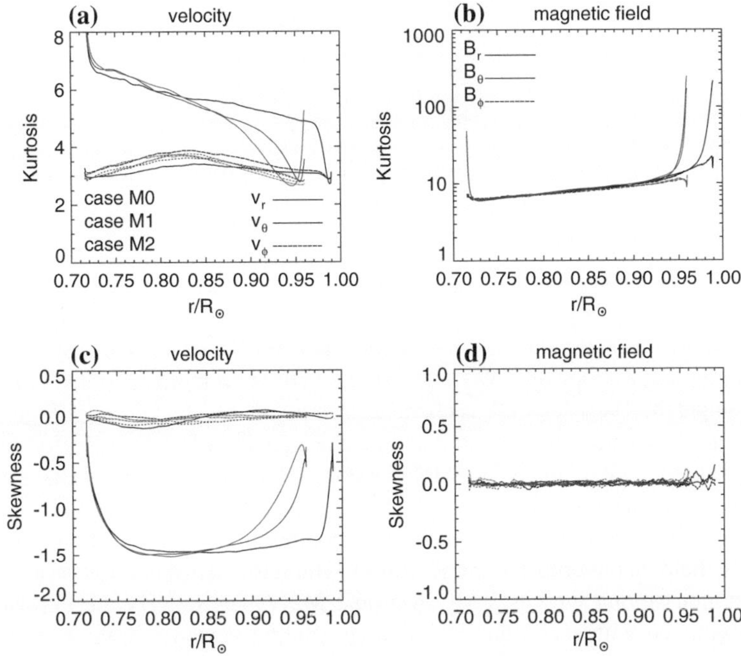

Fig. 3.14 The kurtosis (the forth central moment of the PDF defined in Eq. (3.8): **a, b** and the skewness (the third central moment of the PDF defined in Eq. (3.9): **c, d** for the three components of velocity and magnetic field are shown. The *black, blue* and *red lines* show the results in the case M0, M1, and M2 (Color figure online)

reflects the asymmetry between up- and downflows due to stratification (see also Figs. 3.6 and 3.8). The horizontal velocities (v_θ and v_ϕ) show almost the Gaussian distribution (see also Fig. 3.14). The magnetic fields have high intermittency compared with the velocities (Brandenburg et al. 1996). Despite our asymmetric initial condition for the longitudinal magnetic field $B_\phi = 100\,\mathrm{G}$, the PDFs show a close to symmetric distribution peaked at zero after a sufficiently long temporal evolution. The maximum value of the strength of the magnetic field is around $10^4\,\mathrm{G}$. The PDF for the radial vorticity is similar to that of the magnetic field, i.e., with high intermittency. This is expected from the similarity between the induction and vorticity equations. The horizontal divergence has similarity to the radial vorticity ω_r in the convection zone with high intermittency, while ζ shows the asymmetry near the top boundary similar to the radial velocity v_r. The temperature perturbation has significant asymmetry, which also reflects the asymmetry between the upflow and downflow.

In order to evaluate the influence of the location of the boundary condition and the thickness of the cooling layer, we investigate the moments of the PDF, in particular the kurtosis $\mathcal{K}$ and the skewness $\mathcal{S}$ as

$$\mathcal{K} = \frac{1}{\sigma^4} \int (x - \langle x \rangle)^4 f(x) dx, \tag{3.8}$$

$$\mathcal{S} = \frac{1}{\sigma^3} \int (x - \langle x \rangle)^3 f(x) dx, \tag{3.9}$$

where $f(x)$ is the PDF, x is each variable, and σ is the standard deviation as

$$\sigma = \sqrt{\int (x - \langle x \rangle)^2 f(x) dx}. \tag{3.10}$$

The kurtosis $\mathcal{K}$ and the skewness $\mathcal{S}$ denote the intermittency and the asymmetry of the distribution, respectively. We note that the PDF is normalized as $\int f(x) dx = 1$. For example the Gaussian PDF is characterized by $\mathcal{K} = 3$ and $\mathcal{S} = 0$. Figure 3.14 shows the distribution of the kurtosis and the skewness for the velocity and the magnetic field. As explained above the horizontal velocities (v_r and v_ϕ) have almost the Gaussian distribution, i.e., $\mathcal{K} \sim 3$ and $\mathcal{S} \sim 0$ all over the convection zone. The radial velocity has high intermittency and asymmetry in the convection zone. The magnetic field has intermittency and almost symmetric distribution. These features are common among the cases M0, M1, and M2. Quantitatively the kurtosis and skewness agree with each other in all cases. While the values in the near surface region ($r > 0.85R_\odot$) are influenced by the two factors, the location of the upper boundary and the thickness of the cooling layer, in the deeper region the values converge.

3.3.5 Generation and Transportation of Magnetic Field

In this subsection we investigate the generation and transportation of magnetic field by turbulent thermal convection. The global structure of the mutual interaction between the plasma and the magnetic field is our interest.

Figure 3.4 shows the radial velocity v_r, the radial magnetic field B_r, and the radial vorticity ω_r at the different depth ($r = 0.99R_\odot$: a, b, and c, $r = 0.95R_\odot$: d, e, and f, $r = 0.85R_\odot$: g, h, and i). As introduced in Sect. 3.3.1, there is good coincidence between the downflow and the region with the large amplitude of the radial vorticity. This means that downflows, especially in the deeper region, include most of the turbulent small-scale horizontal motion. We can also see the preferential association of strong magnetic field with downflows and regions with strong vorticity in the constant-depth plane (middle and right columns of Fig. 3.4) and also in the meridional plane (Fig. 3.5). In order to investigate this aspect quantitatively, we take the correlation between the radial velocity v_r and the absolute value of the magnetic field B i.e., $\langle v_r, B \rangle$ in Fig. 3.15a, where our definition of the correlation between quantity A and B is defined as

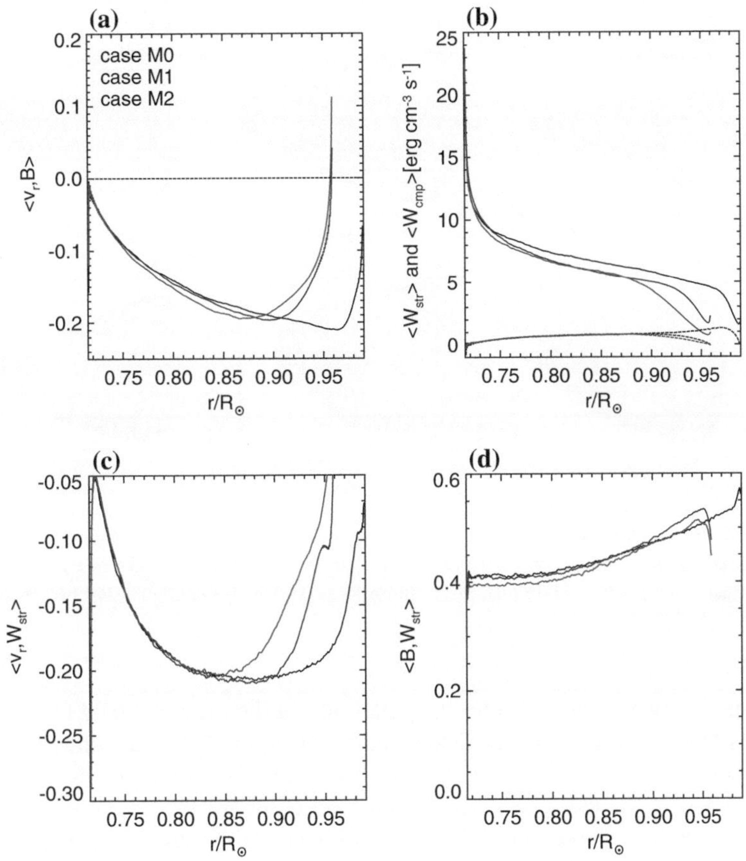

Fig. 3.15 **a** The correlation between the radial velocity and the strength of the magnetic field ($\langle v_r, B\rangle$), **b** the horizontal average of the generation rate of the magnetic energy by the stretching ($\langle W_{\mathrm{str}}\rangle$: *solid lines*) and the compression ($\langle W_{\mathrm{cmp}}\rangle$: *dashed lines*), **c** the correlation between the radial velocity and the generation rate of the magnetic energy ($\langle v_r, W_{\mathrm{str}}\rangle$), **d** the correlation between the strength of the magnetic field and the generation rate of the magnetic energy ($\langle B, W_{\mathrm{str}}\rangle$) are shown

$$\langle A, B\rangle = \frac{\int AB\,dS}{\sqrt{\int A^2\,dS}\sqrt{\int B^2\,dS}}. \tag{3.11}$$

Note that the definition is different from that in Chap. 4. Figure 3.15a shows the negative value in the most of the convection zone. This means that the magnetic field is preferentially found in downflows. This is also seen in the joint PDFs of v_r and B in Fig. 3.16. The figure shows the asymmetric distribution about the $v_r = 0$ axis in the convection zone. Strong magnetic field is more likely to be located in downflow region. We note that a symmetric distribution is found at $r = r_{\mathrm{min}}$, which corresponds to $\langle v_r, B\rangle \sim 0$ in Fig. 3.15a.

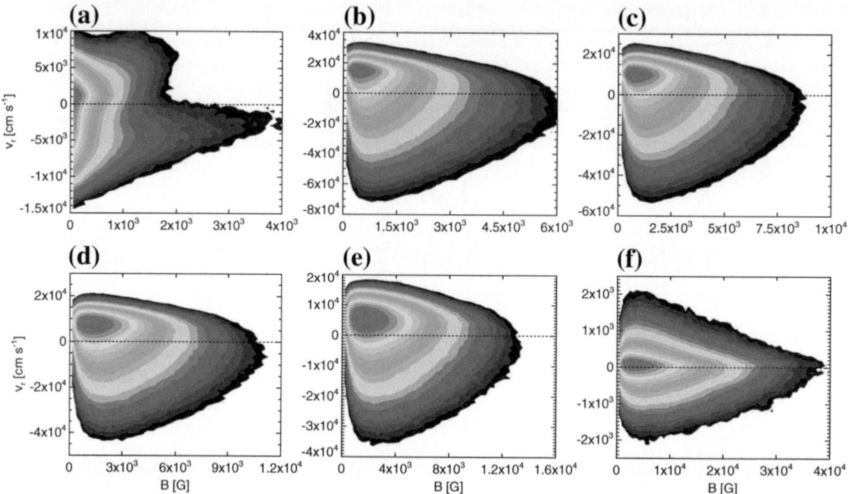

Fig. 3.16 Joint PDFs with the radial velocity (v_r) and the strength of the magnetic field (B) in the case M0 at **a** $r = r_{\max}$, **b** $r = 0.95R_\odot$, **c** $r = 0.90R_\odot$, **d** $r = 0.85R_\odot$, **e** $r = 0.80R_\odot$, **f** $r = r_{\min} = 0.715R_\odot$. The *black lines* shows $v_r = 0$, which distinguish the upflow and the downflow (Color figure online)

There are two possibilities for the preference of the magnetic field to downflow regions. One is that the magnetic field is generated uniformly in space and transported to the downflow region by converging motion. The other is that the magnetic field is generated in the downflow region. In order to answer this question, we define and evaluate the generation rate of magnetic energy by the stretching (W_{str}) and the compression (W_{cmp}) as

$$W_{\mathrm{str}} = \frac{\mathbf{B}}{4\pi} \cdot [(\mathbf{B} \cdot \nabla)\mathbf{v}], \tag{3.12}$$

$$W_{\mathrm{cmp}} = -\frac{B^2}{4\pi}(\nabla \cdot \mathbf{v}). \tag{3.13}$$

The estimations for the horizontal averages, $\langle W_{\mathrm{str}} \rangle$ and $\langle W_{\mathrm{cmp}} \rangle$, are shown in Fig. 3.15b. It is clear that the value of the stretching $\langle W_{\mathrm{str}} \rangle$ is much larger than that of the compression all over the convection zone and the generation of the magnetic field is basically done by the stretching in the turbulent motion. This is also seen by the realistic calculation in the photosphere (Pietarila Graham et al. 2010) in which they suggest that 95 % of the gain of the magnetic energy is done by the stretching. We also take the correlation with the radial velocity v_r and the energy generation rate by the stretching W_{str} (Fig. 3.15c). The distribution of this correlation is similar to that of $\langle v_r, B \rangle$, i.e., the effective stretching prefers the downflow region. Thus we conclude that the reason why the strong magnetic field prefers the downflow region is that the magnetic field is more likely to be generated there. The correlation between B and

W_{str} is also taken. This shows a larger value of (>0.4) throughout the convection zone (Fig. 3.15d). These features discussed above are basically common among the studied cases.

In order to investigate the magnetic field transport in the convection zone, we evaluate the Poynting flux given by:

$$F_{\mathrm{m}} = \frac{c}{4\pi} \left(E_\theta B_\phi - E_\phi B_\theta \right), \tag{3.14}$$

where the electric field is defined as $\mathbf{E} = -(\mathbf{v} \times \mathbf{B})/c$ and c is the speed of light (see Brun et al. 2004). Figure 3.17 shows the horizontally integrated Poynting flux L_{m} as a function of the depth. Since the absolute value ($\sim 10^{31}$ erg s^{-1}) is much smaller than the solar luminosity ($L_\odot = 3.84 \times 10^{33}$ erg s^{-1}), the Poynting flux does not contribute significantly to the total energy flux balance. The Poynting flux is negative in the most of the convection zone, since strong magnetic field is concentrated in downflow regions. This is suggested by previous study as the turbulent pumping effect (Tobias et al. 1998, 2001). Within the convection zone, the flux has a positive value only in the thin layer ($\sim 0.01 R_\odot$) close to the bottom. This is caused by the bounced motion from the bottom boundary. The magnetic energy is transported upward and this provides the magnetic flux in the upflow region. This can be one of the reasons why the absolute values of the correlation $\langle v_r, B \rangle$ is small in the deeper region (Fig. 3.15a) and why the distribution of the joint PDF in the bottom (Fig. 3.16f) is symmetric.

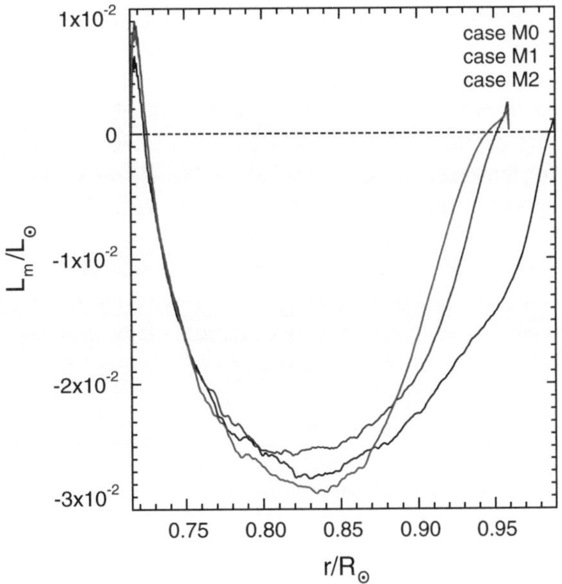

Fig. 3.17 The integrated radial Poynting flux as a function of the depth is shown. The *black, blue* and *red lines* show the results in the case M0, M1, and M2, respectively (Color figure online)

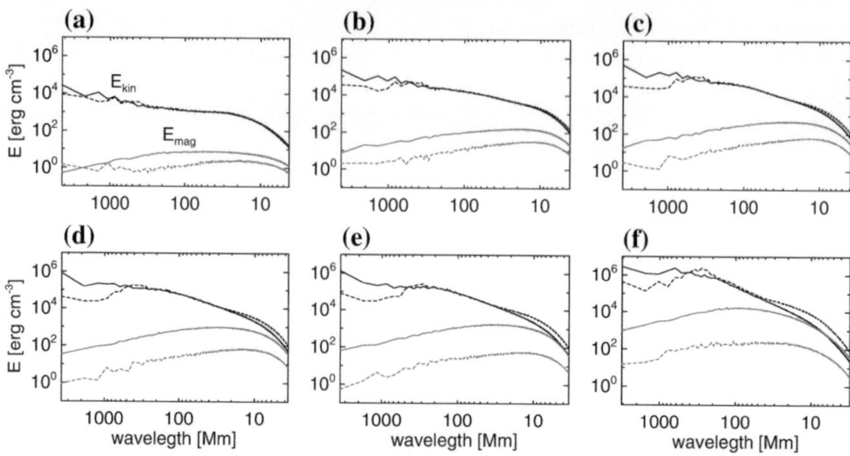

Fig. 3.18 The spectra of the kinetic energy ($E_{\mathrm{kin}} = \rho_0 v^2/2$: *solid black lines*) and the magnetic energy ($E_{\mathrm{mag}} = B^2/(8\pi)$: *solid red lines*) in the case M0. The *dashed black lines* show the kinetic energy in H0, i.e., without the magnetic field. The *dashed red lines* show the magnetic energy at $t = 5.8$ days, **a** r $= r_{\max}$, **b** r $= 0.95R_\odot$, **c** r $= 0.90R_\odot$, **d** r $= 0.85R_\odot$, **e** r $= 0.80R_\odot$, **f** r $= 0.715R_\odot$ (Color figure online)

In the following discussion, we investigate the scale of the magnetic field generated. Figure 3.18 shows the spectra of the kinetic energy (black lines: $E_{\mathrm{kin}} = \rho_0 v^2/2$) and the magnetic energy (red lines: $E_{\mathrm{mag}} = B^2/(8\pi)$) in the case M0 averaged from $t = 173$ day to $t = 237$ day in which the generation of the magnetic field is close to be saturated. The dashed black and red lines show the kinetic energy without the magnetic field and the magnetic energy at $t = 5.8$ days, respectively. In the upper convection zone ($\geq 0.85R_\odot$), the spectra of the magnetic energy peaks at the smallest scale, which is typical for the kinematic phase of a local dynamo. Finding this feature in the saturated phase indicates that the local dynamo is likely not very efficient for our resolved scales (>7 Mm). We see also no indication that the kinetic energy spectrum changed due to the presence of the dynamo near the top of the domain. This situation is different in the lower half of the domain. Figure 3.18e, f show some peak shift of the magnetic energy to larger scales and some feedback on thermal convection, i.e., the kinetic energy is suppressed on the smallest scales. Here the magnetic energy slightly exceeds the kinetic energy near the smallest scales. This shows a more efficient local dynamo can be achieved to some extent for our resolution in lower part of the convection zone ($<0.85R_\odot$).

Figure 3.19 shows the spectra of the horizontal divergence (ζ) and the radial vorticity (ω_r) which show the similar distribution to that of the magnetic field with the peak at small scales. Moffatt (1961) suggested that the power spectrum varies as $D_b(k) \propto k^2 D_v(k)$, when the magnetic field is proportional to ($\nabla \cdot \mathbf{v}$) or ($\nabla \times \mathbf{v}$), where $D_b(k)$ and $D_v(k)$ are magnetic and kinetic spectra. Recently this was confirmed through solar observations in the photosphere using the *Hinode* satellite (Katsukawa and Orozco Suárez 2012). Since in our calculation there is similarity between the

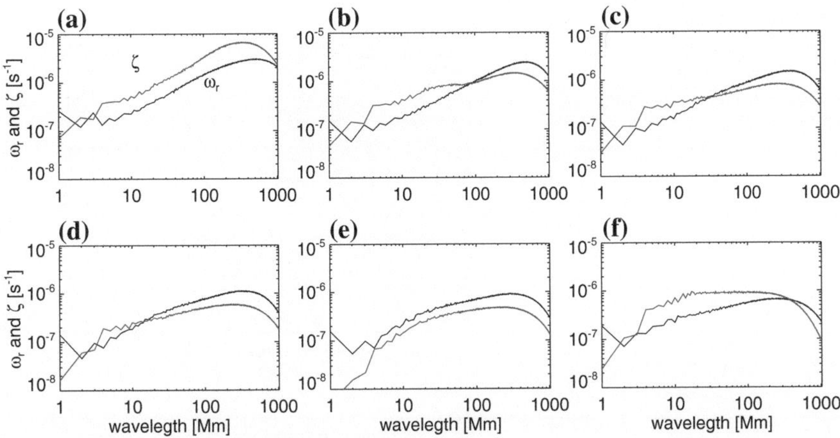

Fig. 3.19 The spectra of the horizontal divergence ζ (*red lines*) and the radial vorticity ω_r (*black lines*) using the result in the case M0, **a** r = r_{max}, **b** r = 0.95R$_\odot$, **c** r = 0.90R$_\odot$, **d** r = 0.85R$_\odot$, **e** r = 0.80R$_\odot$, **f** r = 0.715R$_\odot$ (Color figure online)

magnetic field and the vorticity or the divergence, the magnetic energy peaks at smaller scales than that of the kinetic energy. We note that differences can be seen between shear of the velocity (ω_r and ζ) and the magnetic energy (E_{mag}) at the base of the convection zone (Figs. 3.18f and 3.19f), where the feedback from the magnetic field is stronger.

Even though the numerical resolution does not vary much with depth, the effectiveness of the local dynamo is not expected to be depth independent. There are two reasons for this: The intrinsic convective scale varies with depth and a downward Poynting flux exists almost everywhere in the convection zone. In the upper half of the convection zone the divergence of the Poynting flux provides an energy sink, while at the same time the dynamo is not very efficient on the resolved scale [(similar discussion can be found in Stein et al. (2003)]. In the lower convection zone ($<0.85R_\odot$), the radial gradient of the Poynting flux is negative, thus the magnetic energy is accumulated. At the same time the intrinsic scale of convection is larger and better resolved, leading to a more efficient dynamo. It should be noted that Vögler and Schüssler (2007) found that in even near surface the small-scale convection reproduced with the high resolution has short time scale enough to amplify the magnetic field against the pumping effect.

In order to confirm our idea about the generation and transportation of the magnetic field in our calculation, we evaluate the effective shear for the magnetic field,

$$f_{\text{eff}} = \frac{\langle W_{\text{str}} \rangle}{\langle B^2/(8\pi) \rangle}. \tag{3.15}$$

f_{eff} has the unit of s^{-1} and indicates the time scale of the amplification. Figure 3.20a shows the time evolution of log f_{eff} and indicates that regardless of the phase of the generation of the magnetic field, the larger value of f_{eff} is seen in the upper

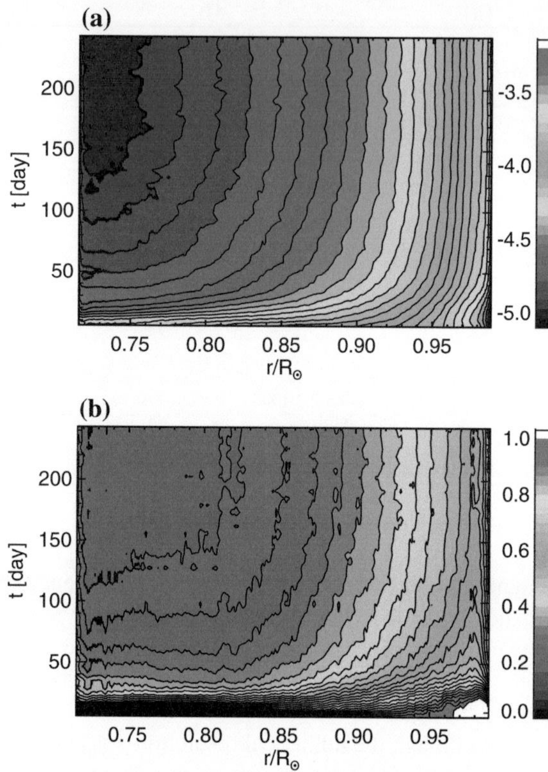

Fig. 3.20 The time evolution of the effective shear **a** log f_{eff} and **b** $f_{\mathrm{eff}}(t)/f_{\mathrm{eff}}(t = 5.8\,\mathrm{day})$

convection, which reflects the short time scale of the thermal convection there. At later times with stronger field, the effective shear f_{eff} is reduced and after $t \sim 150$, it remains constant. Figure 3.20b shows the ratio to the value at $t = 5.8\,\mathrm{day}$, i.e., $f_{\mathrm{eff}}(t)/f_{\mathrm{eff}}(t = 5.8\,\mathrm{day})$. The suppression of the effective shear depends on the depth. It is more suppressed in the deeper region. This might be caused by the feedback of the magnetic field on the velocity seen in Fig. 3.18 (the solid and dashed black lines) as well as a misalignment between shear and magnetic field around the base of the convection zone. Ineffective suppression of f_{eff} in the upper convection zone confirms our idea in previous paragraph. The local dynamo saturates there with little nonlinear feedback (suppression of f_{eff}), since the pumping effect works well in the upper region and the dynamo is not very efficient on the resolved scales to begin with.

Figure 3.21 also shows the variation of dynamo efficiency in the convection zone. Figure 3.21a shows the equipartition magnetic field $B_{\mathrm{eq}} = \sqrt{4\pi v_{\mathrm{rms}}^2}$ and the RMS magnetic field B_{rms} as functions of the depth. The solid and dotted lines show the values at the downflow and upflow regions, respectively. Basically both B_{eq} and B_{rms} increase with the depth and Fig. 3.21b shows an increase of $B_{\mathrm{rms}}/B_{\mathrm{eq}}$ with depth which is caused by the ineffectiveness of the local dynamo in the upper region due to the pumping effect and our insufficient resolution.

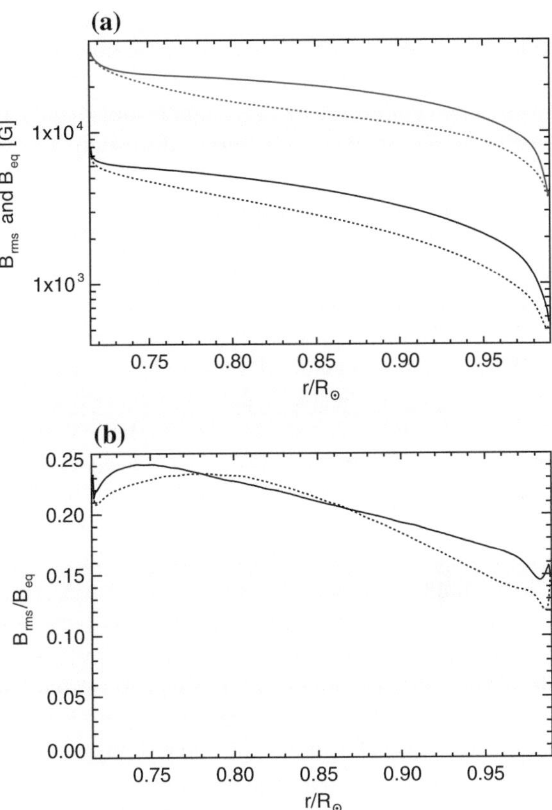

Fig. 3.21 **a** The distribution of the equipartition field B_{eq} (*red lines*) and the RMS magnetic field B_{rms} (*black lines*). **b** The ratio of the RMS magnetic field and the equipartition magnetic field ($B_{\mathrm{rms}}/B_{\mathrm{eq}}$). The *solid* and *dashed* show the values at the downflow and upflow region, respectively (Color figure online)

3.4 Discussion and Summary

We carry out the high-resolution calculations of the solar global convection which resolve the 10 Mm-scale convection smaller than the supergranulation using the RSST (described in Chap. 1). The RSST leads to a simple algorithm and requires only local communication in the parallel computing. In addition, this method has the capability to access the real solar surface without loosing the important physics. This enables us to capture near surface small-scale convection while keeping a global domain. Our main conclusions are listed as follows: (1) Small-scale convection is excited close to the surface ($>0.9R_{\odot}$), when we expand our domain upward to $0.99R_{\odot}$ to capture the near surface layers with small pressure scale heights. (2) In deeper convection zone ($<0.9R_{\odot}$) the convection flow is not influenced by the location of the top boundary and the assumed thickness of the thermal boundary layer. We do not find significant differences in the convective structure and properties of the local dynamo. (3) Using an Large Eddy Simulation (LES) approach we can achieve small scale dynamo action and maintain a field of about 0.15–$0.25 B_{\mathrm{eq}}$ throughout the convection zone. (4) The overall dynamo efficiency varies significantly in the convection zone as a consequence of the downward directed Poynting flux and the depth variation of the intrinsic convective scales. For a fixed numerical resolution the dynamo relevant scales are better resolved in the deeper convection zone and are therefore less affected by numerical diffusivity, i.e., the effective Reynolds numbers are larger.

The conclusion 2 is one of the most important results in this study, since it means that previous calculations (e.g. Miesch et al. 2008) are physically reasonable in the deeper convection zone even if the top boundary condition is placed significantly below the solar surface.

We summarize our simulation results in a schematic shown in Fig. 3.22. Several aspects, in particular with regard to the local dynamo require higher resolution before directly applicable to the solar convection zone. High resolution simulations of a local dynamo in the solar photosphere Vögler and Schüssler (2007) suggest that efficient dynamo action is possible even in the presence of the pumping effect. These simulations use however a grid resolution of about a factor of 100 larger than our setup, which is currently out of reach for global scale convection simulations. Therefore the dynamo RMS field strength of 0.15–$0.25 B_{eq}$ can likely be considered a lower limit.

One issue we cannot address in this study is the problem about the small magnetic Prandtl number ($Pm \sim 10^{-3}$) in the solar convection zone, since we adopt numerical diffusivity which assumes that the magnetic Prandtl number is around unity. Several authors argued that smaller magnetic Prandtl numbers make local dynamos less efficient (e.g. Schekochihin et al. 2004; Boldyrev and Cattaneo 2004). In other word, a small magnetic Prandtl number requires a large magnetic Reynolds number in order to achieve a super-critical dynamo. While this can be a significant problem for numerical simulations with rather moderate Reynolds numbers, this is less likely an issue in the solar convection zone with Reynolds numbers as large as $Rm \sim 10^{11}$ (Brandenburg 2011). Thus, we believe that our approach relying only on the numerical diffusivity can capture the physics of the local dynamo in the solar convection zone.

Fig. 3.22 The schematic view of our calculation. The *arrows* show the convection flow

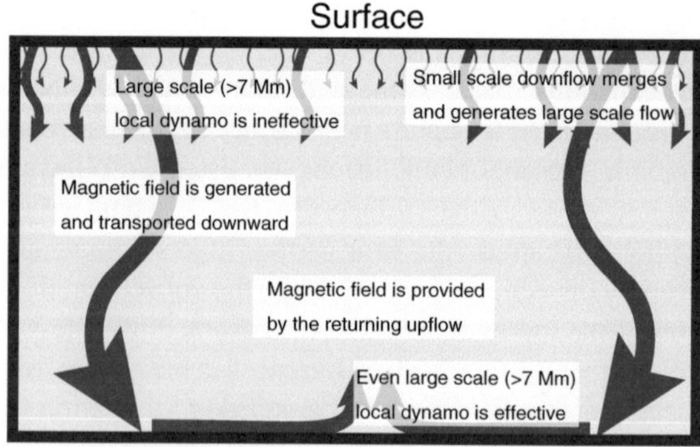

Rotation is not included in this study but is a crucial effect on the generation of the coherent magnetic field. The small-scale convection pattern reproduced in this study may influence the interaction between the convection and the rotation and may improve our previous understanding. This is reported in Chap. 4. By using the RSST, the resolution will be increased following the development of the super computing. The global calculation including the photosphere will be achieved near in the future. Of course the non-uniform grid, such as the nested grid or the adaptive mesh is useful in order to overcome the significant difference in the dynamic scale. These methods are already implemented in our numerical code (Hotta and Yokoyama 2012).

References

S. Boldyrev, F. Cattaneo, Magnetic-field generation in Kolmogorov turbulence. Phys. Rev. Lett. **92**(14), 144501 (2004). doi:10.1103/PhysRevLett.92.144501

A. Brandenburg, Nonlinear small-scale dynamos at low magnetic Prandtl numbers. ApJ **741**, 92 (2011). doi:10.1088/0004-637X/741/2/92

A. Brandenburg, R.L. Jennings, Å. Nordlund, M. Rieutord, R.F. Stein, I. Tuominen, Magnetic structures in a dynamo simulation. J. Fluid Mech. **306**, 325–352 (1996). doi:10.1017/S0022112096001322

A.S. Brun, M.S. Miesch, J. Toomre, Global-scale turbulent convection and magnetic dynamo action in the solar envelope. ApJ **614**, 1073–1098 (2004). doi:10.1086/423835

H. Hotta, T. Yokoyama, Generation of twist on magnetic flux tubes at the base of the solar convection zone. A&A **548**, 74 (2012). doi:10.1051/0004-6361/201220108

H. Hotta, Y. Iida, T. Yokoyama, Estimation of turbulent diffusivity with direct numerical simulation of stellar convection. ApJL **751**, 9 (2012a). doi:10.1088/2041-8205/751/1/L9

H. Hotta, M. Rempel, T. Yokoyama, Y. Iida, Y. Fan, Numerical calculation of convection with reduced speed of sound technique. A&A **539**, 30 (2012b). doi:10.1051/0004-6361/201118268

Y. Katsukawa, D. Orozco Suárez, Power Spectra of velocities and magnetic fields on the solar surface and their dependence on the unsigned magnetic flux density. ApJ **758**, 139 (2012). doi:10.1088/0004-637X/758/2/139

M.S. Miesch, A.S. Brun, M.L. De Rosa, J. Toomre, Structure and evolution of giant cells in global models of solar convection. ApJ **673**, 557–575 (2008). doi:10.1086/523838

M.S. Miesch, J.R. Elliott, J. Toomre, T.L. Clune, G.A. Glatzmaier, P.A. Gilman, Three-dimensional spherical simulations of solar convection. I. Differential rotation and pattern evolution achieved with laminar and turbulent states. ApJ **532**, 593–615 (2000). doi:10.1086/308555.

H.K. Moffatt, The amplification of a weak applied magnetic field by turbulence in fluids of moderate conductivity. J. Fluid Mech. **11**, 625–635 (1961). doi:10.1017/S0022112061000779

J. Pietarila Graham, R. Cameron, M. Schüssler, Turbulent small-scale dynamo action in solar surface simulations. ApJ **714**(2), 1606 (2010). http://stacks.iop.org/0004-637X/714/i=2/a=1606

A.A. Schekochihin, S.C. Cowley, J.L. Maron, J.C. McWilliams, Critical magnetic Prandtl number for small-scale dynamo. Phys. Rev. Lett. **92**(5), 054502 (2004). doi:10.1103/PhysRevLett.92.054502

R.F. Stein, D. Bercik, A. Nordlund, Solar surface magneto-convection, in *Current Theoretical Models and Future High Resolution Solar Observations: Preparing for ATST*. Astronomical Society of the Pacific Conference Series, vol. 286, ed. by A.A. Pevtsov, H. Uitenbroek (2003), p. 121

M. Stix, *The Sun: An Introduction*, 2nd edn. Astronomy and Astrophysics Library (Springer, Berlin, 2004). ISBN: 3-540-20741-4

S.M. Tobias, N.H. Brummell, T.L. Clune, J. Toomre, Pumping of magnetic fields by turbulent penetrative convection. ApJL **502**, 177 (1998). doi:10.1086/311501

S.M. Tobias, N.H. Brummell, T.L. Clune, J. Toomre, Transport and storage of magnetic field by overshooting turbulent compressible convection. ApJ **549**, 1183–1203 (2001). doi:10.1086/319448

A. Vögler, M. Schüssler, A solar surface dynamo. A&A **465**, 43–46 (2007). doi:10.1051/0004-6361:20077253

Chapter 4
Reproduction of Near Surface Shear Layer with Rotation

Abstract We carry out high-resolution calculation of thermal convection in the spherical shell with rotation to reproduce the near surface shear layer (NSSL). It is thought that the NSSL is maintained by thermal convection for small spatial scales and short time scales, which causes a weak rotational influence. The calculation with the RSST succeeds in including such a small scale as well as large-scale convection and the NSSL is reproduced especially at high latitude. The maintenance mechanisms are the following. The Reynolds stress under the weak influence of the rotation transports the angular momentum radially inward. Regarding the dynamical balance on the meridional plane, in the high latitude positive correlation $\langle v'_r v'_\theta \rangle$ is generated by the poleward meridional flow whose amplitude increases with the radius in the NSSL and negative correlation $\langle v'_r v'_\theta \rangle$ is generated by the Coriolis force in the deep convection zone. The force caused by the Reynolds stress compensates the Coriolis force in the NSSL.

Keywords Solar differential rotation $\cdot$ Solar meridional flow $\cdot$ Near-surface shear layer

4.1 Introduction

Howard et al. (1984) compared the rotation rate estimated from Doppler velocity measurements and the tracking of sunspots. It was found that the rotation rate of the sunspots is coherently faster than the Doppler velocity. This indicates that the sunspots are anchored in a faster-rotating deeper layer. Later, the existence of the NSSL was confirmed with global helioseismology (Thompson et al. 2003).

The purposes of this section are: First, to reproduce the NSSL in the numerical calculations and second, to understand the generation and maintenance mechanism of the NSSL.

According to Miesch and Hindman (2011), the mean flows in the convection zone are described by two equations, which are the gyroscopic pumping and the thermal wind balance equations. We discuss the NSSL using these two equations. Gilman and Foukal (1979) suggest that when the convection is not much influenced

© Springer Japan 2015

H. Hotta, *Thermal Convection, Magnetic Field, and Differential Rotation in Solar-type Stars*, Springer Theses, DOI 10.1007/978-4-431-55399-1_4

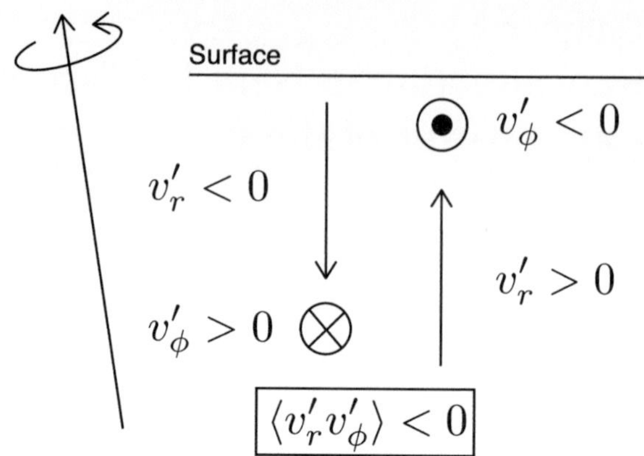

Fig. 4.1 Schematic of the inward angular momentum transport by the radial velocity under the weak influence of rotation

by the rotation, the radial velocity in the thermal convection transports the angular momentum radially inward (Fig. 4.1).When the influence from the rotation is weak and the radial motion conserves the angular momentum, the correlation $\langle v'_r v'_\phi \rangle$ is negative and its corresponding Reynolds stress transports the angular momentum radially inward. Gilman and Foukal (1979) argued that this is the process for generating and maintaining the NSSL. There have been several attempt to reproduce the NSSL with this (De Rosa et al. 2002; Rempel 2005; Brandenburg 2007; Guerrero et al. 2013).

Miesch and Hindman (2011), however, showed that the radially inward angular momentum transport by the Reynolds stress is the only necessary condition and that the thermal wind balance should be considered. When the advection/stretching and baroclinic terms in Eq. (1.15) are neglected, the thermal wind balance equation becomes

$$\frac{\partial \langle \omega_\phi \rangle}{\partial t} = 2r \sin \theta \Omega_0 \frac{\partial \langle \Omega_1 \rangle}{\partial z}, \tag{4.1}$$

where parenthesis $\langle \rangle$ means the zonal average in this chapter. This means that when the radially inward angular momentum transport generates the NSSL especially from mid to high latitudes, i.e., negative $\partial \langle \Omega_1 \rangle / \partial z$, it creates an anticlockwise meridional flow. This meridional flow continues to be accelerated and transport the angular momentum until $\partial \langle \Omega_1 \rangle / \partial z$ becomes zero. This is why the contribution from the baroclinic term and/or the advection/stretching term is required to maintain the non-Taylor-Proudman state $(\partial \langle \Omega_1 \rangle / \partial z \neq 0)$. It is thought that from the middle to the base of the convection zone, the baroclinic term plays an essential role to maintain the non-Taylor-Proudman state, i.e., the conical distribution of the differential rotation and the tachocline (Rempel 2005; Miesch et al. 2006; Brun et al. 2011; Hotta and Yokoyama 2011).

Regarding the NSSL, it is unlikely that the baroclinic term is larger than that in the middle of the convection zone. The advection/stretching term could play an essential role in maintaining the NSSL. In the near surface layer, the convection speed increases and the spatial scale decreases. This causes the large Rossby number, i.e., weak influence of the rotation on the convection flow.

Thus, the reproduction of the NSSL in the numerical calculations requires wide spatial and temporal scales, which must cover the giant cell to supergranulation. Chapter 3 shows that this study succeeds in reproducing convection smaller than the supergranular scale convection in the global simulation domain including the near surface layer with the reduced speed of sound technique without rotation. In this chapter, we include the rotation to reproduce the NSSL in the global convection calculations. The main issue is to clarify the generation and maintenance mechanism of the NSSL in the view of the dynamical balance on the meridional plane as well as the angular momentum transport.

4.2 Model

The numerical model is shown in Chap. 2. We adopt the reduced speed of sound technique (Hotta et al. 2012) and the equation of state including the partial ionization effect for the sun.

We include the effect of rotation with a rate of $\Omega_0/(2\pi) = 413\,\text{nHz}$, which is the solar rotation rate. The radiative diffusivity is 18 times smaller than that calculated with Model S; thus, the inputted luminosity is also 18 times smaller than the solar luminosity. When we use the low viscosity with solar rotation and luminosity, the polar region is accelerated rather than the equator (Fan et al. 2013). The formation of the NSSL, however, requires the small-scale convection pattern which can be achieved with low viscosity. Thus we adopt this radiative diffusivity to decrease the Rossby number in the convection zone in which the acceleration of the equator is reproduced. We assume that the numerically unresolved thermal convection transports substantial energy in the real sun (see the discussion in Chap. 5 in detail).

The top and bottom boundaries are at $r_{\text{max}} = 0.99R_\odot$ and $r_{\text{min}} = 0.715R_\odot$, respectively. In both boundaries, the impenetrate and stress-free boundary conditions are adopted. The resolution is $384(N_r) \times 648(N_\theta) \times 1944(N_\phi) \times 2$ in the Yin-Yang grid.

4.3 Results

The simulated period is 4,500 days. From around $t = 4,000$ day, the distribution of the angular velocity is in statistically steady state in which the angular velocity distribution does not change significantly. In this chapter, the time average is between $t = 4,000$ day and $t = 4,500$ day. To increase the statistical validity, we average the north and south hemispheres considering the symmetry. Figure 4.2 shows a snapshot of the radial velocity v_r at $t = 4,000$ day and selected depths. The white lines show the location of the tangential cylinder $r \sin\theta = r_{\text{min}}$. We can reproduce the supergranulation scale convection at $r = 0.99R_\odot$ without any influence of the

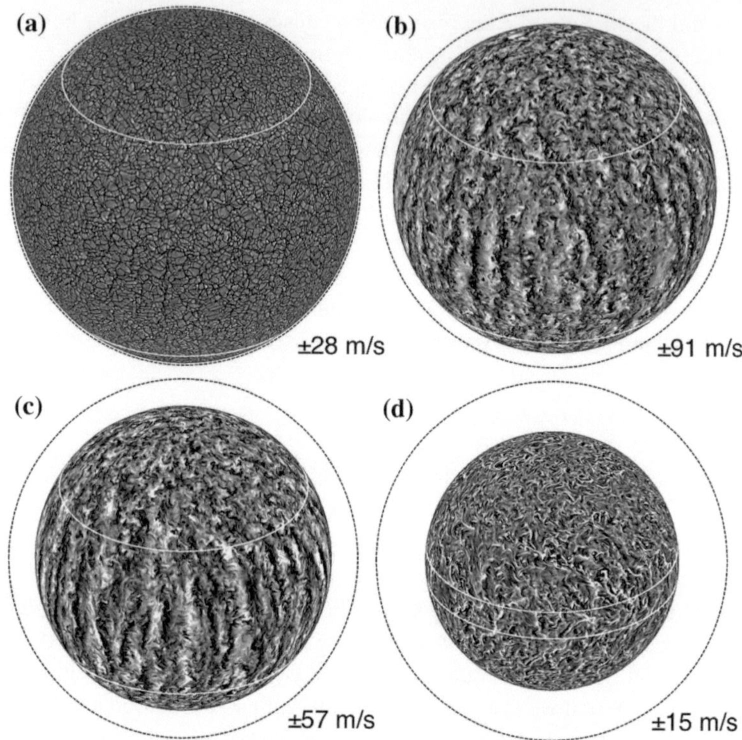

Fig. 4.2 Contour of the radial velocity v_r at **a** $r = 0.99R_\odot$, **b** $r = 0.92R_\odot$, **c** $r = 0.85R_\odot$, and **d** $r = 0.72R_\odot$. The *white lines* show the location of the tangential cylinder $r \sin \theta = r_{\min}$

rotation in which we cannot see any clear alignment of the convection pattern along the rotational axis (the banana cell). At $r = 0.92R_\odot$, the banana cell like feature begins to appear and at $r = 0.85R_\odot$, we can see clear banana cell patterns. In addition, the banana cell pattern is seen outside the tangential cylinder. This dependence of the convection pattern on depth is basically determined by the Rossby number. Figure 4.3a, b show the radial profile of RMS velocity v_{rms} and the Rossby number defined here by Ro $= v_{\mathrm{RMS}}/(2\Omega_0 H_p)$, respectively. Three components of the RMS velocity monotonically increase along the radius. This and the decrease in the pressure scale height H_{p} cause the significant increase of the Rossby number around the surface. Especially above $r = 0.93R_\odot$, the Rossby number exceeds unity indicating weak rotational influence on the convective flow.

Figure 4.4 shows the distribution of the zonally averaged angular velocity $\langle \Omega \rangle/(2\pi)$, where $\Omega = \Omega_1 + \Omega_0$ and $\Omega_1 = v_\phi/(r \sin \theta)$. The NSSL's features are clearly seen, especially in the high latitude ($\theta > 45°$) and low latitude ($\theta < 30°$). In the convection zone at mid to high latitude, the differential rotation is almost in the Taylor-Proudman state ($\partial \langle \Omega_1 \rangle/\partial z \sim 0$). Note that the angular velocity has similar values to the solar one, i.e., 460 and 340 nHz at the equator and the polar regions,

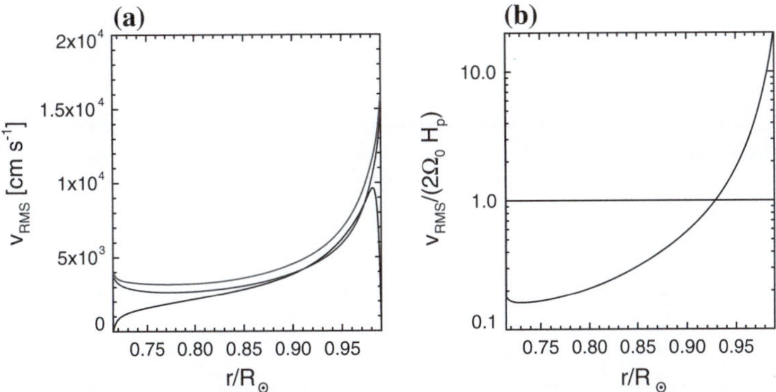

Fig. 4.3 Radial profile of **a** the RMS velocity **b** $v_{\mathrm{RMS}}/(2\Omega_0 H_{\mathrm{p}})$. The *black, blue,* and *red lines* show the radial (v_r), latitudinal (v_θ), and zonal (v_ϕ) values, respectively. The *dashed line* in panel **b** indicates the values at unity (Color figure online)

Fig. 4.4 The angular velocity ($\langle\Omega\rangle/(2\pi)$) on the meridional plane in the unit of nHz

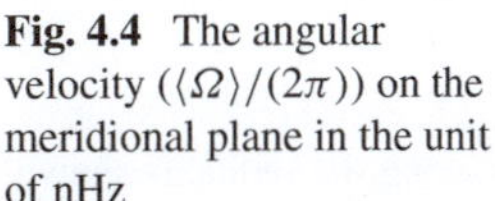

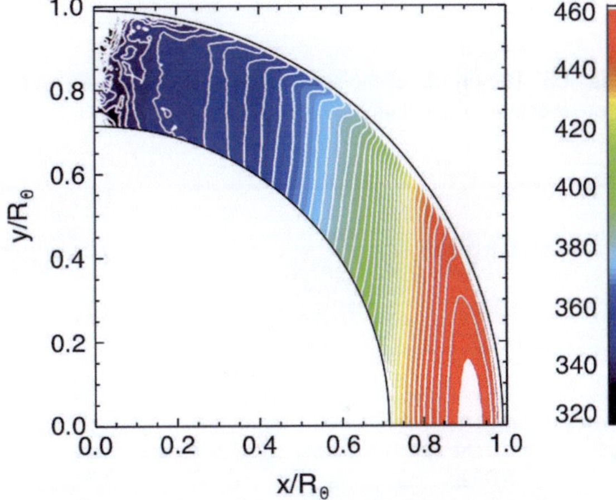

respectively (see Fig. 1.2). Figure 4.5 shows the radial profile of the angular velocity at selected colatitudes. At high latitude ($\theta = 30°$), we can clearly see the decrease of the angular velocity from $r = 0.95R_\odot - 0.985R_\odot$, which is the feature of the NSSL. At $\theta = 45$ and $60°$ the tendency is reversed. The angular velocity increases more steeply than that in the convection zone. At the low latitude, although the rapid increase remains, the decrease from $r = 0.91R_\odot - 0.97R_\odot$ is seen.

Figure 4.6 shows the mean meridional flow. Figure 4.6b clearly shows that in the near surface area ($>0.9R_\odot$), there are a prominent poleward flow and an equatorward meridional flow around the base of the convection zone. Only in the thin layer near the surface at the mid- to low latitude, an equatorward meridional flow is seen. In the convection zone, the multi-cell structure of the meridional flow is generated, which is similar to the recent finding by the local helioseismology (Zhao et al. 2013).

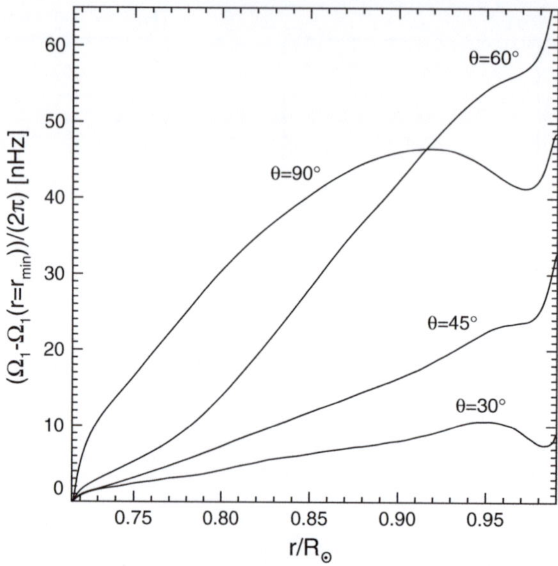

Fig. 4.5 The profile of the angular velocity on the selected latitudes along the radius is shown. Here θ denotes the colatitude

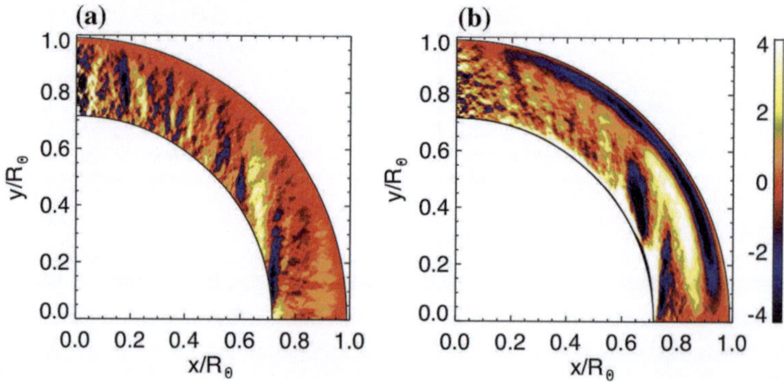

Fig. 4.6 The radial and latitudinal velocity multiplied by the background density ρ_0 and averaged in time and zonal direction. **a** $\rho_0 \langle v_r \rangle$ and **b** $\rho_0 \langle v_\theta \rangle$ in the unit of $\mathrm{g\,cm^{-2}\,s^{-1}}$

Figure 4.7a, b show the correlations between the velocities, i.e., $\langle v_r' v_\phi' \rangle$ and $\langle v_\theta' v_\phi' \rangle$. We note that these correlations are not normalized by the RMS velocity (different from the definition in Chap. 3). The negative correlation of $\langle v_r' v_\phi' \rangle$, speculated by Fig. 4.1, is reproduced, which causes the radially inward angular momentum transport. This negative correlation is not confined to the NSSL. In the convection zone at the low latitude, the positive correlation of $\langle v_\theta' v_\phi' \rangle$ is reproduced, which is generated by the banana cell like feature (Miesch 2005). Figure 4.7c, d show the values $-\nabla \cdot (\rho_0 r \sin\theta \langle \mathbf{v}_m' v_\phi' \rangle)$, and $\rho_0 \langle \mathbf{v}_m \rangle \cdot \nabla \langle \mathscr{L} \rangle$, respectively, i.e. the balance of the

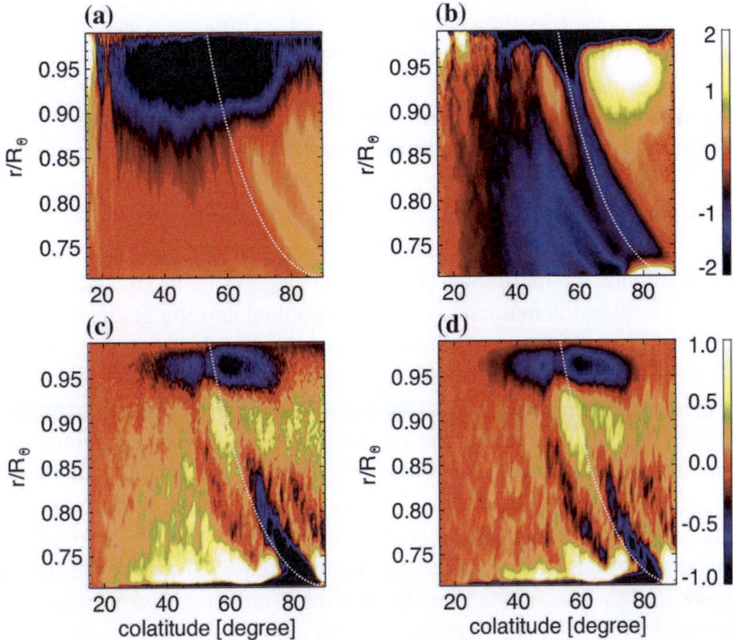

Fig. 4.7 The values **a** $\langle v_r' v_\phi' \rangle$, **b** $\langle v_\theta' v_\phi' \rangle$ in the unit of $10^6\,\mathrm{cm^2\,s^{-2}}$ and **c** $-\nabla \cdot (\rho_0 r \sin\theta \langle \mathbf{v}_m' v_\phi' \rangle)$, and **d** $\rho_0 \langle \mathbf{v}_m \rangle \cdot \nabla \langle \mathscr{L} \rangle$ in the unit of $10^6\,\mathrm{g\,cm^{-1}\,s^{-2}}$ are shown on the meridional plane

angular momentum transport (see Eq. 1.14). The angular momentum transports by the Reynolds stress and the mean meridional flow are roughly balanced in the convection zone. The difference between these indicates the longer-time evolution of the differential rotation as well as the effectiveness of the artificial viscosity which is not estimated in this study.

As introduced in Sect. 4.1, the discussion regarding the thermal wind balance is required to understand the maintenance mechanism of the NSSL in addition to the angular momentum transport shown in Fig. 4.7. We divide the contribution of the dynamical balance on the meridional plane as

$$-\mathscr{T} = \mathscr{B} + \tilde{\mathscr{C}} + \mathscr{C}', \tag{4.2}$$

where

$$\mathscr{T} = 2r(\sin\theta)\Omega_0 \frac{\partial \langle \Omega_1 \rangle}{\partial z}, \tag{4.3}$$

$$\mathscr{B} = \frac{g}{\rho_0 r} \left(\frac{\partial \rho}{\partial s} \right)_p \frac{\partial \langle s_1 \rangle}{\partial \theta}. \tag{4.4}$$

The term $\mathscr{T}$ is caused by the Coriolis force and contributes when the differential rotation is away from the Taylor-Proudman state ($\partial\langle\Omega_1\rangle/\partial z \neq 0$). The term $\mathscr{B}$ is caused by the pressure gradient and the buoyancy (baroclinic term) and is effective when the latitudinal entropy gradient is generated. The detailed form of the $\mathscr{C}'$ and $\tilde{\mathscr{C}}$ are found in Appendix. These two are caused by the momentum transport on the meridional plane. $\mathscr{C}'$ and $\tilde{\mathscr{C}}$ are contribution by the mean flow ($\langle v_r\rangle$ and $\langle v_\theta\rangle$) and the nonaxisymmetric flow (v'_r and v'_θ), respectively. Figure 4.8 shows the distribution of (a) $-\mathscr{T}$, (b) $\mathscr{B}$, (c) $\tilde{\mathscr{C}}$, and (d) $\mathscr{C}'$. According to the distribution of $-\mathscr{T}$ in Fig. 4.8a, we divide the meridional plane to four regions (I, II, III, and IV as shown in Fig. 4.8a). Region I is maintained by the latitudinal entropy gradient $\mathscr{B}$ from the middle to the bottom of the convection zone (see panel b). In the other regions (II, III, and IV), the deviation from the Taylor-Proudman state cannot be compensated by the entropy gradient. The mean flow ($\tilde{\mathscr{C}}$: Fig. 4.8c) has negligible role almost everywhere including the NSSL. We see that the contribution from the non-axisymmetric flow ($\mathscr{C}'$: Fig. 4.8d) almost compensates the term $-\mathscr{T}$ at the regions II, III, and IV. We note that the region III is not well balanced between $-\mathscr{T}$ and $\mathscr{C}'$, indicating long-time evolution or the lack of sampling in averaging especially for $\mathscr{C}'$. To investigate the origin of the distribution of $\mathscr{C}'$, which can maintain the NSSL, we divide the term $\mathscr{C}'$ to three as $\mathscr{C}' = \mathscr{C}'_d + \mathscr{C}'_\theta + \mathscr{C}'_r$. Their detailed forms are found in Appendix. The term $\mathscr{C}'_d$ is caused by the diagonal component of the momentum flux F_{rr}, $F_{\theta\theta}$, and $F_{\phi\phi}$.

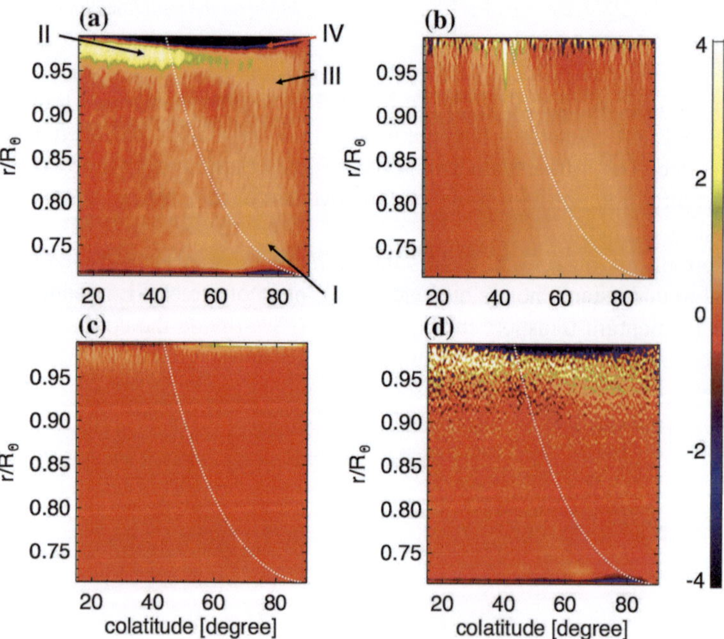

Fig. 4.8 The values **a** $-\mathscr{T}$, **b** $\mathscr{B}$, **c** $\tilde{\mathscr{C}}$, and **d** $\mathscr{C}'$ are shown in the unit of $10^{-12}\,\mathrm{s}^{-2}$ are shown on the meridional plane

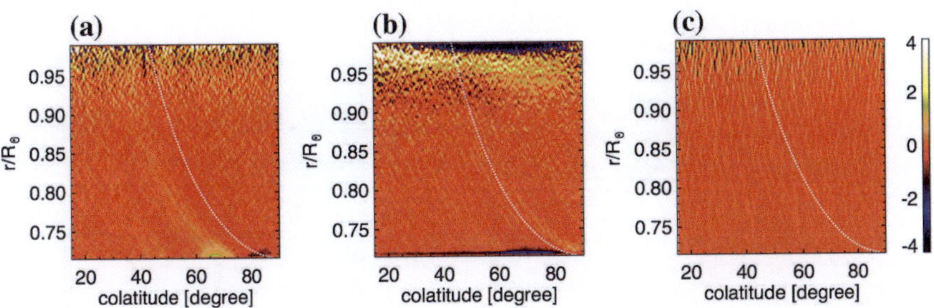

Fig. 4.9 The values **a** $\mathscr{C}_{\mathrm{d}}'$, **b** $\mathscr{C}_{\theta}'$, and **c** $\mathscr{C}_{r}'$ in the unit of $10^{-12}\,\mathrm{s}^{-2}$ are shown on the meridional plane

The terms $\mathscr{C}_{\theta}'$ and $\mathscr{C}_{r}'$ are caused by the nondiagonal components of the momentum flux $F_{r\theta}$. The difference between these two terms is explained as follows: the term $\mathscr{C}_{\theta}'$ ($\mathscr{C}_{r}'$) is caused by the transport of the latitudinal momentum $\rho_0 v_\theta$ (radial momentum $\rho_0 v_r$) in the radial (latitudinal) direction. Figure 4.9 shows the distribution of (a) $\mathscr{C}_{\mathrm{d}}'$, (b) $\mathscr{C}_{\theta}'$, and (c) $\mathscr{C}_{r}'$. The diagonal term $\mathscr{C}_{\mathrm{d}}'$ has contribution to some degree and the contribution from the term $\mathscr{C}_{r}'$ is negligible. The essential contribution is by the term $\mathscr{C}_{\theta}'$, i.e., the transport of the latitudinal momentum in the radial direction.

Next we investigate the origin of the distribution of the $\mathscr{C}_{\theta}'$ by estimating the value

$$D_{\theta(n)}' = -\frac{1}{\rho_0}\left[\frac{1}{r^2}\frac{\partial}{\partial r}(r^2 F_{r\theta}') - \frac{F_{\theta r}'}{r}\right], \tag{4.5}$$

and the correlation $\langle v_r' v_\theta' \rangle$, where $F_{ij}' = \rho_0 \langle v_i' v_j' \rangle$ and

$$\mathscr{C}_{\theta}' = \frac{1}{r}\frac{\partial}{\partial r}(r D_{\theta(n)}'). \tag{4.6}$$

Figure 4.10 shows (a) $D_{\theta(n)}$ and (b) $\langle v_r' v_\theta' \rangle$. $D_{\theta(n)}$ indicates the direction of the latitudinal force by the momentum transport. The direction is equatorward (poleward) in the top (bottom) of the NSSL in the high latitude (Region II). This force compensates the Coriolis force there. The origin of this force by the momentum transport is shown in Fig. 4.10b. Around the high latitude NSSL, the positive correlation is reproduced. In the high latitude deeper convection zone, the correlation is negative. This increases (decreases) the latitudinal momentum in the upper (lower) part of the NSSL (Fig. 4.10a). The positive and negative correlations are the essential maintenance mechanism of the NSSL in the high latitude.

Before discussing on the generation of the correlation in the high-latitude NSSL, let us mention the low-latitude deeper layer feature. From the mid- to the low-latitude, the negative correlation is reproduced in the near surface layer. The positive correlation is generated in the middle of the convection zone at the lower latitude. This positive correlation is generated by the banana cell. In the banana cell region, the

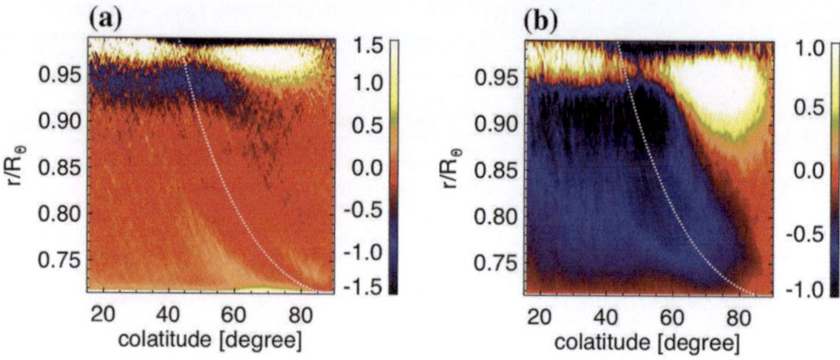

Fig. 4.10 The values **a** $D_{\theta(\mathrm{n})}$ in the unit of $10^{-3}\,\mathrm{cm\,s^{-2}}$ and **b** $\langle v'_r v'_\theta \rangle$ in the unit of $10^6\,\mathrm{cm^2\,s^{-2}}$ are shown on the meridional plane

Coriolis force is effective and the zonal velocity v_ϕ significantly generates the radial and latitudinal velocities. When both radial and latitudinal velocities are generated by the Coriolis force, the correlation $\langle v'_r v'_\theta \rangle$ can be positive. In the NSSL, however, the banana cell is unlikely to exist under the high Rossby number condition. This means that a positive correlation $\langle v'_r v'_\theta \rangle$ in the high latitude NSSL is generated by different mechanism(s).

From here, we discuss the origin of correlation between the velocities. Basically, the existence of the mean flow is a cause of the anisotropic correlated flow. The other terms are likely to cause the correlation to be zero. We retain the terms that can generate a positive or negative correlation as follows:

$$\frac{\partial v'_r}{\partial t} = -\frac{v'_\theta}{r}\frac{\partial \langle v_r \rangle}{\partial \theta} + 2v'_\phi \Omega_0 \sin\theta + [\cdots], \tag{4.7}$$

$$\frac{\partial v'_\theta}{\partial t} = -v'_r \frac{\partial \langle v_\theta \rangle}{\partial r} + 2v'_\phi \Omega_0 \cos\theta + [\cdots], \tag{4.8}$$

$$\frac{\partial v'_\phi}{\partial t} = -2v'_r \Omega_0 \sin\theta - 2v'_\theta \Omega_0 \cos\theta + [\cdots]. \tag{4.9}$$

The first term in each of the Eqs. (4.7) and (4.8) is that due to the mean meridional flow that is the most important element in this discussion. In this discussion, we focus on the correlation between v'_r and v'_θ. When the typical time scale is estimated to be $\tau = H_p/v_{\mathrm{RMS}}$, we obtain the relation

$$v'_\phi \sim -2\tau v'_r \Omega_0 \sin\theta - 2\tau v'_\theta \Omega_0 \cos\theta, \tag{4.10}$$

from Eq. (4.9). We substitute this relation to Eqs. (4.7) and (4.8), and only retain the terms that can generate a nonzero correlation between v'_r and v'_θ:

$$\frac{\partial v'_r}{\partial t} = [\cdots] - \frac{v'_\theta}{r}\frac{\partial \langle v_r \rangle}{\partial \theta} - 2v'_\theta \tau \Omega_0^2 \sin(2\theta), \tag{4.11}$$

$$\frac{\partial v_\theta'}{\partial t} = [\cdots] - v_r' \frac{\partial \langle v_\theta \rangle}{\partial r} - 2v_r'\tau\Omega_0^2\sin(2\theta). \qquad (4.12)$$

This means that the terms from the Coriolis force (i.e. the last term in each equation) generate a negative correlation between v_r' and v_θ'. This is understood by thinking that when the Coriolis force is strong, the fluid is likely to move along the rotation axis. The sign of the correlation by the mean flow depends on the sign of $\partial\langle v_r\rangle/(r\partial\theta)$ and $\partial\langle v_\theta\rangle/\partial r$. Figure 4.11 shows the distribution of (a) $\partial\langle v_r\rangle/(r\partial\theta)$, and (b) $\partial\langle v_\theta\rangle/\partial r$. It is clear that the contribution from the term related to $\partial\langle v_r\rangle/(r\partial\theta)$ is small compared with the term of $\partial\langle v_\theta\rangle/\partial r$. Interestingly we find that the negative value of $\partial\langle v_\theta\rangle/\partial r$ in region II and the positive value is found in region IV. Only when $\partial\langle v_\theta\rangle/\partial r$ has negative value, the correlation $\langle v_r'v_\theta'\rangle$ can have a positive value in Eq. (4.12). The effectiveness of the generation of the positive correlation by the mean meridional flow over the Coriolis force can be estimated as follows:

$$\mathcal{M} = -\frac{\partial\langle v_\theta\rangle/\partial r}{2\tau\Omega_0^2\sin(2\theta)} \sim -\frac{1}{\sin(2\theta)\Omega_0}\frac{\partial\langle v_\theta\rangle}{\partial r}\mathrm{Ro}. \qquad (4.13)$$

Because the positive correlation is found at $\theta = 20\text{–}40°$, we estimate $\sin(2\theta) \sim 0.5$. $\mathrm{Ro} = v_{\mathrm{RMS}}/(2\Omega_0 H_p) \sim 4$ which is estimated in Fig. 4.3 at the base of the NSSL and $|\partial\langle v_\theta\rangle/\partial r| \sim -5 \times 10^{-7}\,\mathrm{s}^{-1}$ at this region (around $r = 0.96R_\odot$). $\mathcal{M}$ at the base of NSSL is around 1.5. This shows that the generation of the positive correlation by the mean poleward flow begins to be effective in the base of the NSSL. When the value $\partial\langle v_\theta\rangle/\partial r$ is positive both terms of the meridional flow and the Coriolis force generate the negative correlation. This cannot generate the solar-like NSSL even under the large Rossby number situation (region IV).

Regarding the low-latitude NSSL (region III), the positive correlation $\langle v_r'v_\theta'\rangle$ generated by the banana cell have a role with some contribution of the poleward meridional flow there (Fig. 4.10b). Around the tangential cylinder (white line) the effect by the banana cell and the meridional flow is ineffective and the correlation

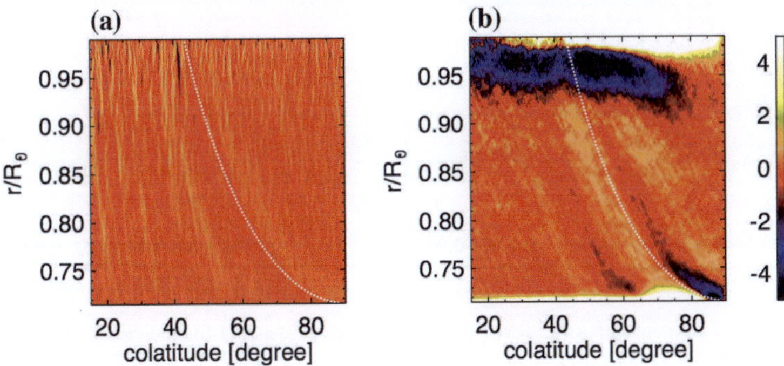

Fig. 4.11 The values **a** $\partial\langle v_r\rangle/(r\partial\theta)$, and **b** $\partial\langle v_\theta\rangle/\partial r$ in the unit of $10^{-7}\,\mathrm{s}^{-1}$ are shown on the meridional plane

$\langle v'_r v'_\theta \rangle$ is negative. In the boundary of the effective and ineffective layers of these mechanisms, the fluid is accelerated poleward, which compensates the Coriolis force in the low-latitude NSSL.

4.4 Calculation with High Rotation Rate and Solar Luminosity

In this section, we show an additional calculation with 2.4 times of the solar rotation ($\Omega_0/2\pi = 989\,\mathrm{nHz}$) and with the solar luminosity. In this calculation we adopt rather low resolution $256(N_r) \times 432(N_\theta) \times 1{,}296(N_\phi) \times 2$. Even with this resolution, convection pattern around upper boundary can have supergranulation scale. The radiative diffusivity calculated in Model S is adopted in this calculation, i.e., the solar luminosity is imposed from the bottom boundary. The main purpose of this section is to investigate the relation between imposed luminosity, rotation rate, and obtained NSSL pattern. Figure 4.12a, b show the RMS velocity and the Rossby number, respectively with the same manner as Fig. 4.3. Figure 4.12a shows that the RMS velocity increases from the result in the previous section with factor of 2.7 due to the increased luminosity. Since the rotation rate is larger in this case than the previous, we have similar distribution of the Rossby number (Figs. 4.3b and 4.12b). Figure 4.13a, b shows the mean meridional flow, i.e., $\rho_0 \langle v_r \rangle$ and $\rho_0 \langle v_\theta \rangle$, respectively. The differential rotation is shown in Fig. 4.14. Comparison with Figs. 4.4 and 4.6 indicates that the essential features do not change and the NSSL is established also with this setting. We conduct same analysis as the previous section and conclude that the NSSL in this setting is also maintained by the same mechanism.

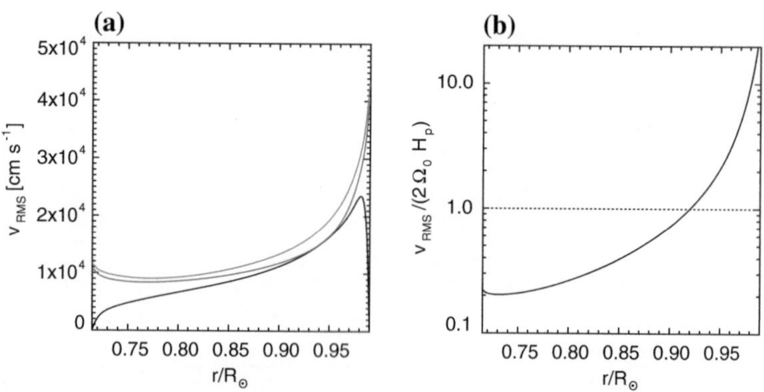

Fig. 4.12 Radial profile of **a** the RMS velocity, **b** $v_\mathrm{RMS}/(2\Omega_0 H_\mathrm{p})$ for the calculation with 2.4× of the solar rotation and with 1.0× of the solar luminosity. The *black, blue,* and *red lines* show the radial (v_r), latitudinal (v_θ), and zonal (v_ϕ) values, respectively. The *dashed line* in panel **b** indicates the values at unity (Color figure online)

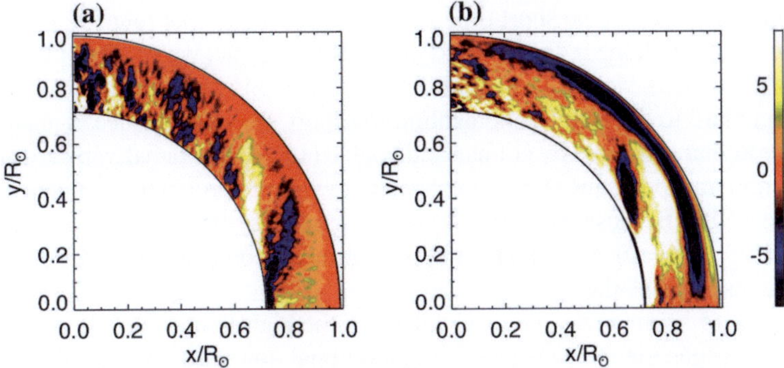

Fig. 4.13 The radial and latitudinal velocity multiplied by the background density ρ_0 and averaged in time and zonal direction for the calculation with $2.4\times$ of the solar rotation and with $1.0\times$ of the solar luminosity. **a** $\rho_0\langle v_r\rangle$ and **b** $\rho_0\langle v_\theta\rangle$ in the unit of $\mathrm{g\,cm^{-2}\,s^{-1}}$

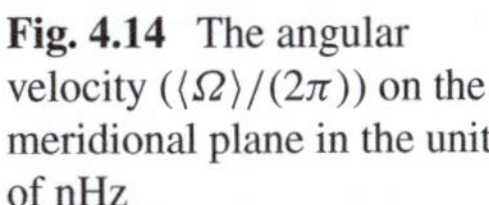

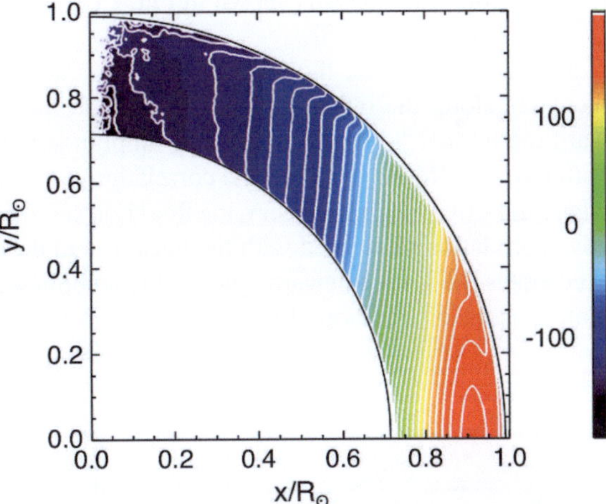

Fig. 4.14 The angular velocity ($\langle\Omega\rangle/(2\pi)$) on the meridional plane in the unit of nHz

This result shows that even with different imposed luminosity, i.e., different convection velocity, and different rotation rate, the NSSL pattern is obtained with a similar distribution of the Rossby number. This means that when we investigate different star with different luminosity and rotation rate, the Rossby number is an essential value to understand the differential rotation.

4.5 Summary and Discussion

We conduct the high-resolution calculation of the thermal convection in the spherical shell with rotation in the highly stratified layer for the purpose of reproducing the NSSL. It is thought that the NSSL is maintained by the thermal convection with the small spatial scale and the short time scale which causes weak rotational influence. By calculation with the reduced speed of sound technique, we succeed in including such small-scale as well as large-scale convection and in reproducing the NSSL.

With regard to the angular momentum transport, the maintenance mechanism is the same as that suggested by Gilman and Foukal (1979). The radially inward angular momentum transport caused by the convection with weak rotational influence maintains the NSSL. Because the NSSL is significantly away from the Taylor-Proudman state ($\partial\langle\Omega_1\rangle/\partial z \neq 0$), a mechanism(s) is required to compensate the Coriolis force which tends to break the NSSL.

Figure 4.15 summarizes the mechanism. In the high-latitude NSSL, the positive correlation is generated by the poleward meridional flow whose amplitude increases with the radius ($\partial\langle v_\theta\rangle/\partial r < 0$). In the deeper convection zone, the Coriolis force produces the negative correlation $\langle v_r' v_\theta'\rangle$ with the alignment of the flow along the rotational axis. The positive correlation $\langle v_r' v_\theta'\rangle$ in the NSSL transports latitudinal momentum upward, and thus, the top (bottom) of the NSSL is accelerated equatorward (poleward). This is shown in Fig. 4.15b. This acceleration compensates the Coriolis force in the NSSL.

In the low latitude, the banana cell generates the positive correlation, which increases along the radius and accelerates the fluid poleward. When the equatorward meridional flow with increasing amplitude ($\partial\langle v_\theta\rangle/\partial r > 0$) is effective, i.e., with the large Rossby number, the correlation $\langle v_r' v_\theta'\rangle$ becomes negative. At the layer where this effect begins to occur, the fluid is accelerated equatorward. Then, the negative correlation multiplied with the background density becomes zero, which then accelerates the fluid poleward again. This complicated transport of the momentum maintains the distribution of the NSSL at the low latitude in our calculation.

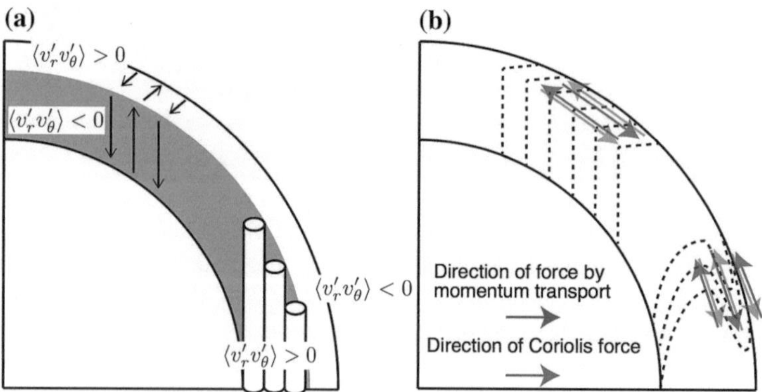

Fig. 4.15 The summary of our findings is in the schematic. The panel **a** shows the distribution of the correlation $\langle v_r' v_\theta'\rangle$. The *gray* area indicates the strong influence of the rotation. The panel **b** shows the force balance on the meridional plane. The *gray* and *red arrows* show the direction of the Coriolis force and the force by the momentum transport. The *dashed lines* are the contour line of the angular velocity (Color figure online)

Most important finding in this study is that the radial gradient of the latitudinal meridional flow, i.e., $\partial \langle v_\theta \rangle / \partial r$, has an essential role for the maintenance of the NSSL. In this study, the equatorward meridional flow in the very near surface layer in the mid and low latitude is generated. Although the origin of this equatorward meridional flow is unknown, this type of feature is seen in the previous study (Miesch et al. 2008). The equatorward meridional flow is mainly generated where the difference in the angular momentum transport by the Reynolds stress and the mean meridional flow is seen (see Fig. 4.6c, d). This indicates the contribution of the artificial viscosity on this issue. The distribution of the NSSL especially in the low latitude should be confirmed with higher-resolution calculations in the future. In the sun, it is thought that the meridional flow is basically poleward from low to high latitude with increasing its amplitude (Zhao et al. 2013). This can generate a positive correlation $\langle v'_r v'_\theta \rangle$ at all the latitudes. It can be thought that this is the reason why the sun has the NSSL at all latitudes.

References

A. Brandenburg, Near-surface shear layer dynamics, in *IAU Symposium*, vol. 239, ed. by F. Kupka, I. Roxburgh, K.L. Chan (2007), pp. 457–466. doi:10.1017/S1743921307000919

A.S. Brun, M.S. Miesch, J. Toomre, Modeling the dynamical coupling of solar convection with the radiative interior. ApJ **742**, 79 (2011). doi:10.1088/0004637X/742/2/79

M.L. De Rosa, P.A. Gilman, J. Toomre, Solar multiscale convection and rotation gradients studied in shallow spherical shells. ApJ **581**, 1356–1374 (2002). doi:10.1086/344295

Y. Fan, N. Featherstone, F. Fang, Three-dimensional MHD simulations of emerging active region flux in a turbulent rotating solar convective envelope: the numerical model and initial results (ArXiv e-prints, 2013)

P.A. Gilman, P.V. Foukal, Angular velocity gradients in the solar convection zone. ApJ **229**, 1179–1185 (1979). doi:10.1086/157052

G. Guerrero, P.K. Smolarkiewicz, A. Kosovichev, N. Mansour, Solar differential rotation: hints to reproduce a near-surface shear layer in global simulations, in *IAU Symposium*, vol. 294, ed. by A.G. Kosovichev, E. de Gouveia Dal Pino, Y. Yan (2013), pp. 417–425. doi:10.1017/S1743921313002858

H. Hotta, T. Yokoyama, Modeling of differential rotation in rapidly rotating solar-type stars. ApJ **740**, 12 (2011). doi:10.1088/0004-637X/740/1/12

H. Hotta, M. Rempel, T. Yokoyama, Y. Iida, Y. Fan, Numerical calculation of convection with reduced speed of sound technique. A&A **539**, 30 (2012). doi:10.1051/0004-6361/201118268

R. Howard, P.I. Gilman, P.A. Gilman, Rotation of the sun measured from Mount Wilson white-light images. ApJ **283**, 373–384 (1984). doi:10.1086/162315

M.S. Miesch, Large-scale dynamics of the convection zone and tachocline. Living Rev. Sol. Phys. **2**, 1 (2005)

M.S. Miesch, B.W. Hindman, Gyroscopic pumping in the solar near-surface shear layer. ApJ **743**, 79 (2011). doi:10.1088/0004-637X/743/1/79

M.S. Miesch, A.S. Brun, J. Toomre, Solar differential rotation influenced by latitudinal entropy variations in the tachocline. ApJ **641**, 618–625 (2006). doi:10.1086/499621

M.S. Miesch, A.S. Brun, M.L. De Rosa, J. Toomre, Structure and evolution of giant cells in global models of solar convection. ApJ **673**, 557–575 (2008). doi:10.1086/523838

M. Rempel, Solar differential rotation and meridional flow: the role of a subadiabatic tachocline for the Taylor-Proudman balance. ApJ **622**, 1320–1332 (2005). doi:10.1086/428282

M.J. Thompson, J. Christensen-Dalsgaard, M.S. Miesch, J. Toomre, The internal rotation of the sun. ARA&A **41**, 599–643 (2003). doi:10.1146/annurev.astro.41.011802.094848

J. Zhao, R.S. Bogart, A.G. Kosovichev, T.L. Duvall Jr., T. Hartlep, Detection of equatorward meridional flow and evidence of double-cell meridional circulation inside the Sun. ApJL **774**, 29 (2013). doi:10.1088/2041-8205/774/2/L29

Chapter 5
Concluding Remarks

Abstract The achievements and the important findings in this thesis are summarized in this chapter. In this thesis we focus on the interaction of the small-scale and large-scale convection in the solar and stellar interior. Although this requires huge numerical resources and technical innovation, we succeed in simultaneously reproducing them. Then the achievements of this thesis provide significant understanding in the convection zone, such as the convection structure itself, the generation and transport of the magnetic field and the maintenance mechanism of the differential rotation. Finally we show remaining problems for the future.

Keywords Solar pole's acceleration problem · Weak solar convection · Solar sear-surface region

5.1 Summary of Thesis

5.1.1 Achievements

1. We significantly improve the ability of the numerical calculation of the solar and stellar global convection.
   Taking advantage of the reduced speed of sound technique as well as some other numerical techniques introduced in Chap. 2, we develop the efficient numerical code for the solar global flow. The numerical code efficiently scales up to 10^5 cores and shows a good performance (14 % to the peak and 3×10^5 grid update/core/s). This is able to cover the broad spatial and temporal range of the plasma in the solar and stellar convection zone.
2. 10 Mm-scale convection is reproduced in the solar global convection for the first time.
   In the previous study, on account of lack of the resolution and large diffusivity, only the $\sim$50 Mm-scale convection is achieved around their upper boundary (Miesch et al. 2008). Because we succeed in increasing the resolution and adopting the higher upper boundary, we can establish the 10 Mm-scale convection even in the global computation domain for the sun. This is reported in Chap. 3.

© Springer Japan 2015

H. Hotta, *Thermal Convection, Magnetic Field, and Differential Rotation in Solar-type Stars*, Springer Theses, DOI 10.1007/978-4-431-55399-1_5

3. The NSSL is reproduced for the first time. The reproduction of the NSSL requires simultaneous establishment of the small- and large-scale convection. This means that reproduction requires a large number of time integration and the grid points. This challenging purpose is achieved in this thesis taking advantage of the good efficiency of our developed numerical code. This is reported in Chap. 4.

5.1.2 Findings

1. The small-scale convection generated near the surface layer influences the relatively shallower layer in the convection zone ($>0.9R_\odot$). In the deeper layer ($<0.9R_\odot$) the influence is negligible.
2. The magnetic field preferentially appears in the downflow, because strong magnetic field is likely generated there. This causes the downward Poynting flux in the convection zone. Around the base of the convection zone ($<0.85R_\odot$), the magnetic energy is accumulated by the Poynting flux. The local dynamo even in the large scale is effective.
3. The NSSL is maintained by the Reynolds stress. The convective motion under the small influence of the rotation transports the angular momentum radially downward. This causes the poleward meridional flow with increasing the amplitude along the radius in the upper convection zone. This meridional flow again causes a positive correlation $\langle v'_r v'_\theta \rangle$ in the near surface layer, which then transports the latitudinal momentum radially upward. In the deeper convection zone, the correlation $\langle v'_r v'_\theta \rangle$ is negative owing to the Coriolis force, which transports the latitudinal momentum radially downward. The force by this momentum transport maintains the NSSL in our calculation. We argue that this is a possible mechanism for the solar case.

5.2 Remaining Problems and Future Work

5.2.1 Comparison with Helioseismology

Most recent observation by Hanasoge et al. (2012) using the local helioseismology technique estimates the internal structure of the thermal convection. They constraint the amplitude of the horizontal velocity associated with the solar convection for the modes with horizontal harmonics $\ell < 60$. The flow speed in this scale is substantially smaller than the result with the ASH code (by two orders of magnitude at maximum). The spectra in this thesis show similar amplitude to that of the ASH code results even with higher resolution and the higher upper boundary. In the current situation, both results of numerical simulation and helioseismology are under debate. A sophisticated comparison between them should be continued to understand the real structure of the solar global convection.

5.2.2 Proper Reproduction of Solar Differential Rotation

There is another problem still remaining for the differential rotation reported from the ASH code group. They investigated the dependence of the differential rotation on the thermal diffusivity and the kinetic viscosity, which is believed to be very small in the real solar interior. They found that if the thermal diffusivity is small ($<3 \times 10^{12}$ cm^2 s^{-1}), the polar region is accelerated, i.e. anti-solar differential rotation is obtained. When the thermal diffusivity is small, the entropy gradient near the surface becomes large and the generated thermal convection velocity becomes larger. This means the Rossby number becomes large. Because in the sun, the Rossby number is thought to be around unity, i.e., the effects of the Coriolis force and the convection are similar, the slight change in the convection velocity can cause significant change in the correlation terms ($\langle v_r' v_\phi' \rangle$ and $\langle v_\theta' v_\phi' \rangle$), the meridional flow and the differential rotation.

This problem might be caused by the limited resolution in the solar global convection simulations. The turbulence in the thermal convection is thought to have the power law distribution from the injection scale to the Kolmogorov dissipation scale. Although the unresolved scale convection can also transport the energy, we ignore them. As a consequence, our resolved scale must transport too much energy and must have too large an amplitude in the velocity. The improvement in this thesis for increasing the resolution will contribute to this fundamental issue in the future.

References

S.M. Hanasoge, T.L. Duvall, K.R. Sreenivasan, Anomalously weak solar convection. Proc. Natl. Acad. Sci. **109**, 11928–11932 (2012). doi:10.1073/pnas.1206570109

M.S. Miesch, A.S. Brun, M.L. De Rosa, J. Toomre, Structure and evolution of giant cells in global models of solar convection. ApJ **673**, 557–575 (2008). doi:10.1086/523838

Appendix

Dynamical Balance on Meridional Plane

In the appendix, we derive the equations for the dynamical balance on the meridional plane. We start with the hydrodynamic equation with the Coriolis force used and with the equation of continuity with the RSST, we obtain,

$$\frac{\partial \mathbf{v}}{\partial t} + \frac{\xi^2 \mathbf{v}}{\rho_0} \frac{\partial \rho_1}{\partial t} = -\frac{1}{\rho_0} \nabla \cdot (\rho_0 \mathbf{v}\mathbf{v}) - \frac{\nabla p_1 + \rho_1 g \mathbf{e_r}}{\rho_0} + 2\mathbf{v} \times \mathbf{\Omega}_0. \tag{A.1}$$

The zonal component of the curl of the second and third terms in the left hand side of Eq. (A.1) are shown in Chap. 1. We define the momentum flux on the meridional plane as follows:

$$\langle F_{ij} \rangle = \tilde{F}_{ij} + F'_{ij}, \tag{A.2}$$

$$\tilde{F}_{ij} = \rho_0 \langle v_i \rangle \langle v_j \rangle, \tag{A.3}$$

$$F'_{ij} = \rho_0 \langle v'_i v'_j \rangle, \tag{A.4}$$

where i and j correspond to r, θ, and ϕ. For this definition, we divide the velocity as $v_i = \langle v_i \rangle + v'_i$. Then the divergence of the fluxes are divided to several terms as follows:

$$\mathbf{D} = -\frac{1}{\rho_0} \nabla \cdot \mathbf{F} = D_r \mathbf{e_r} + D_\theta \mathbf{e}_\theta, \tag{A.5}$$

$$D_r = D_{r(\mathrm{d})} + D_{r(\mathrm{n})}, \tag{A.6}$$

$$D_\theta = D_{\theta(\mathrm{d})} + D_{\theta(\mathrm{n})}, \tag{A.7}$$

$$D_{r(\mathrm{d})} = -\frac{1}{\rho_0} \left[\frac{1}{r^2} \frac{\partial}{\partial r} (r^2 F_{rr}) - \frac{F_{\theta\theta} + F_{\phi\phi}}{r} \right], \tag{A.8}$$

© Springer Japan 2015
H. Hotta, *Thermal Convection, Magnetic Field, and Differential Rotation in Solar-type Stars*, Springer Theses, DOI 10.1007/978-4-431-55399-1

$$D_{r(n)} = -\frac{1}{\rho_0 r \sin\theta} \frac{\partial}{\partial\theta} (\sin\theta \, F_{\theta r}), \tag{A.9}$$

$$D_{\theta(d)} = -\frac{1}{\rho_0} \left[\frac{1}{r \sin\theta} \frac{\partial}{\partial\theta} (\sin\theta \, F_{\theta\theta}) - \frac{\cot\theta \, F_{\phi\phi}}{r} \right], \tag{A.10}$$

$$D_{\theta(n)} = -\frac{1}{\rho_0} \left[\frac{1}{r^2} \frac{\partial}{\partial r} (r^2 F_{r\theta}) - \frac{F_{\theta r}}{r} \right]. \tag{A.11}$$

We use the notation of $\tilde{D} = D(\tilde{F})$ and $D' = D(F')$. Then the zonal component of the curl of the $\langle \mathbf{D} \rangle$ is also divided to several terms as follows:

$$\mathscr{C} = (\nabla \times \langle \mathbf{D} \rangle)_\phi = \mathscr{C}_r + \mathscr{C}_\theta + \mathscr{C}_d. \tag{A.12}$$

Then each term is divided $\mathscr{C}_i = \tilde{\mathscr{C}}_i + \mathscr{C}'_i$, where i corresponds to r, θ, d. The terms are

$$\tilde{\mathscr{C}}_r = -\frac{1}{r} \frac{\partial \tilde{D}_{r(n)}}{\partial\theta}, \tag{A.13}$$

$$\mathscr{C}'_r = -\frac{1}{r} \frac{\partial D'_{r(n)}}{\partial\theta}, \tag{A.14}$$

$$\tilde{\mathscr{C}}_\theta = \frac{1}{r} \frac{\partial}{\partial r} \left(r \tilde{D}_{\theta(n)} \right), \tag{A.15}$$

$$\mathscr{C}'_\theta = \frac{1}{r} \frac{\partial}{\partial r} \left(r D'_{\theta(n)} \right), \tag{A.16}$$

$$\tilde{\mathscr{C}}_d = \frac{1}{r} \frac{\partial}{\partial r} \left(r \tilde{D}_{\theta(d)} \right) - \frac{1}{r} \frac{\partial \tilde{D}_{r(d)}}{\partial\theta}, \tag{A.17}$$

$$\mathscr{C}'_d = \frac{1}{r} \frac{\partial}{\partial r} \left(r D'_{\theta(d)} \right) - \frac{1}{r} \frac{\partial D'_{r(d)}}{\partial\theta}. \tag{A.18}$$

Then the Eq. (A.1) is averaged in time and zonal direction We assume $\partial/\partial t = 0$ then the equation of the balance is obtained as follows:

$$-\mathscr{T} = \mathscr{B} + \mathscr{C}, \tag{A.19}$$

where

$$\mathscr{T} = 2r \sin\theta \, \Omega_0 \frac{\partial \langle \Omega_1 \rangle}{\partial z}, \tag{A.20}$$

$$\mathscr{B} = \frac{g}{\rho_0 r} \left(\frac{\partial \rho}{\partial s} \right)_p \frac{\partial \langle s_1 \rangle}{\partial \theta}. \tag{A.21}$$

Curriculum Vitae

Dr. Hideyuki Hotta received his B.S. degree in 2009 and his PhD in 2014 at University of Tokyo in Japan. He started his post-doc at High Altitude Observatory in USA as a scientific visitor supported by Japan Society for the Promotion of Science from 2014. He is studying solar and stellar global convection, mean field, and dynamo with using High-performance computing.

© Springer Japan 2015
H. Hotta, *Thermal Convection, Magnetic Field, and Differential Rotation in Solar-type Stars*, Springer Theses, DOI 10.1007/978-4-431-55399-1